Steven Hawking

THE WELL-DRESSED APE

HANNAH HOLMES

THE WELL-DRESSED APE

A NATURAL HISTORY OF OURSELVES

ATLANTIC BOOKS
LONDON

First published in the United States of America in 2008 by Random House, an imprint of The Random House Publishing Group, a division of Random House, Inc., New York.

First published in Great Britain in hardback in 2009 by Atlantic Books, an imprint of Grove Atlantic Ltd.

10 9 8 7 6 5 4 3 2 1

A CIP catalogue record for this book is available from the British Library.

978 1 84887 037 6

Typeset in Janson by Ben Cracknell Studios
Printed in Great Britain by the MPG Books Group

Atlantic Books
An imprint of Grove Atlantic Ltd
Ormond House
26–27 Boswell Street
London
WC1N 3JZ

www.atlantic-books.co.uk

IN MEMORY OF MY DAD

DR. P. K. HOLMES

A BIOLOGIST IN THE LARGEST SENSE OF THE WORD

CONTENTS

INTRODUCTION

I am one of those people with a reputation for being a "natural" with children. Because I produced none of my own, my friends often make the observation with an air of puzzlement, after I've beguiled their offspring out of a sulk or into a game quieter than hurling pot lids.

The honest explanation has seemed too impolite to share: Of course I'm fluent in child. I've spent my whole life around wild animals.

I grew up on a small farm in Maine, but livestock was only part of the picture. Besides the domestic cows, pigs, horses, and chickens, we always had a few wilder species, too. In a town without a veterinarian, my parents, both of them biologists, stood out as the experts in broken wings and orphaned babies. Things arrived on the doorstep. Among the family photos is a portrait of me as an infant with a sparrow named Maybe on my head. Maybe, named for her odds of survival, matured and moved outdoors, but would hop up and down on the door latch to come in for a visit. Wowl was an elegant little screech owl who would ride our shoulders, his claws piercing through our shirts while he preened the tangles in our hair. There was a chipmunk named Tammy who lived in a cage in the bathroom, opposite the guinea pigs. They were joined at some point by a flying squirrel, the softest animal I've ever known. Sleepy in the daylight, he'd bumble down inside our collars to ball up in the hammock where shirt tucked into jeans. At night he exercised by sprinting across the floor and up the wall, then kicking off to sail across the room. An ominous silence settled over the bathroom community when a great horned owl arrived for rehab. Confined to an old barrel to keep him from flapping a broken wing, he clattered his scimitar beak whenever humans invaded his territory to take a pee.

We children soon built our own menagerie. The robin who fell from his nest would wait on my shoulder when I shucked corn, eager to peck worms from the cobs. My brother kidnapped a European starling from a hollow tree, rearing it in a berry basket by the window. My sister filched a warm seagull egg from its nest, and the bird and the girl spent a storybook summer together before the adolescent bird would no longer fly home from the shore when she called. Twice I raised orphaned raccoons, which are endearing little bundles of spring steel until they grow up and become large, scary bundles of spring steel. So I know animals. Animal behavior and body language are so legible to me that it demands no conscious effort to interpret it.

This is why I don't find children baffling. They are young animals, unrefined in their instincts and impulses. If an animal is shy, I don't gaze or grab at it, because those gestures are predatory. Instead, I avert my eyes and display something enticing. To avoid frightening the young human who has approached, it's essential to project positive feelings. When a horse detects the stiffening of a fearful rider, the horse tenses because it has evolved to respect any indication of danger. Inversely, a fearful horse can be soothed by a rider who is at ease. And so it is with the young human: He monitors other humans for hesitations, signs of doubt, signs of danger. I try not to embody any. Thus, by exploiting an animal's instincts, it's possible to manipulate its behavior to suit yourself.

Of course, there are differences between children and chipmunks. For one thing, human young are experts at learning. And once they learn they're being manipulated, they often rebel. Second, as humans mature, our enormous brains allow for enormous differences in behavior from one of us to the next. When you wish to manipulate the behavior of an adult human, it becomes more efficient to reason with the animal than to exploit its basic instincts.

Despite the way my early experience with animals has deepened my understanding of humans, I grew up believing a bold line separated my species from all others: There are animals, and there are humans. After all, in my everyday world, the complexity of human behavior underscores our uniqueness and distracts us from the universal traits that unite humans with all other creatures.

But then, for a previous book, I spent a year studying the small ecosystem of my backyard. I got to know my local squirrels and crows, worms and ants, and learned how they all interact with their environment. It wasn't until the end of that book that I circled around to the animal that is me. What are the differences, I began to wonder, between children and chipmunks? I mean, what are the real, biological, brain-ological, immutable differences? And more intriguing, what are the real, biological, brain-ological, immutable commonalities? It was then that I realized I'd never seen a biological fact sheet on the species we call *Homo sapiens.* And that struck me as strange.

Whenever biologists discover a new animal it's their custom to crank the creature through a factual sausage grinder, producing tidy links of information. With academic detachment they tabulate the number of legs and teeth, note food preferences, and characterize habits of reproduction. A porcupine, for instance, emerges with a fact sheet something like this:

> PHYSICAL DESCRIPTION: *This is a fifteen-pound mammal with big teeth and little eyes. Specialized hairs on the back puncture the mouth of predators.*
> HABITAT: *The animal prefers to feed in treetops but will also browse on the ground. It rests in rock burrows when available.*
> RANGE: *North America, including the tundras of Canada and Alaska.*
> BEHAVIOR: *The animal is nocturnal and mostly solitary. Contrary to myth he cannot hurl his quills; in fact, he can become stuck to his victim when the quills refuse to separate.*
> REPRODUCTION: *Precarious.*

And so on, addressing the animal's perceptive senses, communication, diet, environmental impacts, and predators. Every species chugs through the same machinery, emerging as a standardized profile. The fact sheet is a handy way to summarize an animal's place in the web of life.

I've read hundreds of these, describing everything from the three-toed sloth to the nine-banded armadillo and the thirteen-lined ground squirrel. But I've never encountered a full description of the two-legged ape. We *Homo sapiens*, so eager to describe the rest of the world, have been chary about committing our own natural history to paper.

This seems unfortunate. For one thing, it reinforces the notion that we're not normal animals. It lends the impression that we're too wonderful to summarize; that although the giraffe can be corralled in paragraphs, the human cannot. That's unfair to other species. On the flip side, it suggests we're misfits, as animals go. It lends the impression that we're not worthy to take our place beside the gemsbok and the gorilla; that we are excluded from the brotherhood of mammals. This is unfair to my species.

It also seems unnecessarily dour. What could be more fun than describing the human, after all? What color would you consider the animal to be? Regarding diet, is there anything on Earth that we humans won't put in our mouths? As for communication, does my smile or my outstretched palm send the same message as a chimpanzee's version of those gestures? Can the human mate with any other species, the way donkeys can mate with horses, or lions with tigers?

A proper description of the species will answer these questions, and some larger ones, too: Who are we, animally speaking? Sure, we're clever—but compared to what? Yes, we're obsessed with mating, but any more or less than other animals? And our males behave quite differently from our females—but is that unusual? Are humans apex predators like lions or bears, or do we have to watch our backs like gazelles and rabbits? Can we survive as high in the mountains as mountain goats? And if we can, how many square miles does each human require? Sure, we communicate a great deal, but so do parrots and prairie dogs. Our behavior is tremendously tool centered, but the list of other creatures who make and wield implements is growing steadily as we watch them more closely.

Happily, the human (and only the human) delights in analyzing itself. Giraffe nor Gila monster will spend time with chin in hand, watching her neighbors and wondering. But humans analyze ourselves for the fun of it. We, and only we, want to know where the child and the chipmunk overlap, and where they diverge.

THE WELL-DRESSED APE

1
QUICK AS A CRICKET:
PHYSICAL DESCRIPTION

Homo sapiens is a mammal that, uncharacteristically, travels on two legs, leaving the forelimbs free for other tasks. Although the usual gait is a walk, a rare feature in the animal's neck permits it to run at considerable speed for astounding distances. The fur, black in most individuals, is largely restricted to the head and the junctions of limbs and trunk, although a fine pelage does cover the rest of the body. The human skin is usually some shade of brown, but can be pink in the least sunny parts of its range. Eye color is also brown in most humans, but among individuals with reduced skin pigmentation eyes may be hazel, green, or blue.

The human's dentition is typical of an omnivore, excepting its strangely diminutive canine teeth. (In fact, this animal is surprisingly ill-equipped for physical aggression or defense.) Due to prolonged isolation of breeding populations, humans have evolved numerous morphs, or races. "Pygmies" are the smallest of these, standing less than five feet tall. Northern Europeans are the largest, with the Dutch male averaging five feet ten inches.

Sex differences are pronounced. The female carries two permanently enlarged mammary glands high on the chest. As in most mammals, the male has nipples but cannot nurse young. The sexes also differ in height, fat storage, and fur distribution. Although the human bears a resemblance to the other great apes, a careful observer will note that the human demonstrates a stronger inclination to tamper with its skin and fur, and that its excursions into trees are normally clumsy, and often injurious.

HIGHLY VARIABLE IN SIZE

In daily life I pass for a tall, blond human of the northern persuasion. In this regard I am not an ideal subject for a study of the human animal. The majority of humans have darker coloration, are physically adapted to warmer climates, and are somewhat stunted by poor diet. But the stock I have to work with is limited.

There's me. My genes spiral back through time in a series of tall, pale primates, in a trail that goes cold in the British Isles a few centuries ago. Even the randy dentist who hopped the fence to sire my father's mother was a tall blondie, according to the one photo that remained after he was driven from his small Maine town.

And there's my mate, although he's not particularly eager to shuck his clothes and undergo prolonged dissection. Besides, he's not much different, in terms of where his DNA ripened. A strand of his emerged further south than my family's, but not by much. His hide is a bit more olive, and his fur is darker, but he's still decidedly European. And his offspring are a pale pair, too. We're a northern-bred bunch, although the children's hair and eye colors do approach the norm.

But that's us only in daily life. In the biological view, we're perfectly human. The animal I see in my mirror is indisputably human, once you get past the blanched skin. For better or worse, I'm an acceptable specimen of *Homo sapiens*. And right now, the "worse" is what I'm seeing.

I close the shades and cast the clothing aside. My initial survey is disconcerting. I'm naked like a dolphin, but upright like an owl. I'm padded with fat, but have the legs of a stork. The bulb of my cranium swells over my eyes, not behind them. And I cannot deny that the animal before me has been—as though I weren't strange enough—painting itself.

Peculiar beastie. Well, the human animal is no glittering peacock, or tyger burning bright. But by considering my anatomy one feature at a time, I expect the funny elements of my body will add up to something serious. Nature does not produce freaks just for the fun of it. There is always a method to the madness.

In the gloom I give myself a closer look. Not all humans experience anxiety about revealing their hairy parts and their assorted lumps and bumps. Lucky them. My culture has strict ideas about which parts should be revealed and how large or small those parts should be. I take these strictures to heart. Even with the shades drawn I'm nervous, standing here au naturel. This particular human isn't at ease even on the nude beaches of California. Parts bounce. Things rub. And even when the other humans there don't give me a second glance, their first glance is enough. Gimme garments. But this is science. My task is to view the human body as I would a squirrel's body or a walrus's—no, like a squirrel's body: with candid dispassion.

VARIATION IN STATURE

So from the top, which is high. Since my legs lengthened in my early teens, I have been at the bottom of the cheerleading pyramid, the back of the chorus risers, the end of the line from short to tall. Again, not normal. But there is a good explanation or two. New research is taking old beliefs about human size and standing them on their head.

Early students of human anatomy believed some groups of humans were genetically tall, while others were genetically short. A dearth of data handicapped those early students, blinding them to a central fact of human height: It changes. I don't mean that your height and my height change. Rather, the average height of entire groups of humans grows and shrinks with the passing decades. Since that timescale is too short to be the work of evolution, some portion of human height must be determined by the human environment.

A vivid case of this is found a few centuries back, when European humans migrated to the shores of North America. The race that was already in that territory, now known as American Indians, were among the tallest humans on the planet. Males of the Cheyenne tribe averaged five feet ten inches. Why? They enjoyed low population density. They ate a high-protein, low-fat diet of wild greens and buffalo. Their water must have been fairly clean, keeping digestive ailments and parasites to a minimum. They were, in a word, nourished. The migrating Europeans were not.

I wish I knew the height of my ancestors when they arrived, some of them in the late 1600s. Most of them were well funded and well fed, so perhaps they weren't as tiny as the average. Columbus's countrymen (the males) averaged five feet six inches. It's lucky for those European migrants that they had mastered metal tools, because if the stunted specimens had arrived unarmed, they would have been laughed right back home by the strapping locals.

Even if my successful, blond ancestors were tall, I doubt they were the giants we are today.

A wondrous thing happened to those stunted European migrants: Once they took root, they shot up like weeds. Two centuries after Captain John Smith (five four, perhaps) dropped anchor at Jamestown in 1607, the average U.S. male had gained two or three inches, depending on whose data you use. After the European migrants learned how to grow food in North America, the offspring of new immigrants would gain an inch in a single generation. And today they may gain more than that. Some recent migrants from Guatemala have averaged an increase of 2.2 inches in a single generation. In every case, the cause is the same: better food, fewer stunting diseases.

And we're still gaining. National food shortages during World War II meant both my parents lived through a period of food rationing as children. If that means they were undernourished you wouldn't know it from their height. Dad made it to six three. My brother gained an inch on him, though, suggesting that something held Dad down. Mom, who lost some height to the old-fashioned polio virus, is another long drink of water. My sister and I didn't gain any altitude on her, even though we ate a farm-grown diet and evaded polio. The length of my skeleton may represent the full genetic potential of my DNA.

So the height of the human animal waxes and wanes with the quality of its food supply. The case in point: Recent decades have brought improvements to the formerly fetid territory of Europe, from whence my ancestors fled. Dutch males are now the tallest humans on the planet, averaging five feet ten inches. Why Holland? With socialized medicine and financial support for the poor, the Dutch culture ensures that the offspring of all parents get sufficient calories and sufficient protection from disease. Thus each individual comes closer to reaching his potential. (They may still have a ways to grow. Some scholars expect the Dutch to rise another four inches before they hit their genetic ceiling.) By contrast, in my culture food and medicine are shared less equally, with many of our young failing to find healthful food and medical care. Thus poor humans on average are an inch shorter than rich ones in my culture, depressing the national average. And obviously, in realms like Guatemala and Bangladesh, where the distribution of food and safe shelter is more uneven, human height is severely stunted. Sometimes, a discrete event is enough to push human height up or down. Europeans shrank during the Little Ice Age of the 1600s when crops failed. Japanese humans lost height in the hungry years after World War II. We would need a few generations of ideal nutrition for all the world's humans before we could see what the underlying genetic height of various populations might be. For now we can say that the height of the human animal varies by race, from four feet eight inches in male Efe Pygmies of Zaire to five feet ten inches in the Dutch.

The human female is, on average, a few inches shorter than the male. This is a telling detail. Often, a sharp size difference between sexes indicates a violent relationship between males. The goal of a male gorilla, for instance, is to wage war against all other males to control a group of females. The losers don't get to breed. As a result of their high-stakes mating system, male gorillas have evolved to nearly twice the size of females. Chimpanzees live in looser groups with more mate mixing, and the males are just 20 to 30 percent bigger than the females. And among humans, who are somewhat monogamous (we'll test that in chapter 7), males are just 10 or 15 percent larger than females.

When an animal species displays this sort of "size dimorphism," it's usually the male who's biggest. Females usually opt to stop growing and start reproducing, whereas males are forced to grow, in a sort of arms race with other males. But occasionally size dimorphism swings in the other direction. The female spotted hyena is one of the most butch chicks in the animal kingdom. This girl is generally 15 percent bigger than the males, who cower in her presence. Her modified clitoris is about the size of the male's penis. She pees through it, mates through it, and even gives birth through it. (I can hear you wincing, gentlemen.) Although rare, this reversed size dimorphism can evolve if the female is the defender of a territory; or if her method of reproducing demands room for storage of eggs or fat; or, as is suspected of the hyena, a mutation of the hormonal system blitzes a female with extra testosterone.

In some animals, size dimorphism can go to grotesque extremes. Among many spider species the males are so small they could be mistaken for prey. The minuscule males of some orb-weaver spiders have evolved to strum a lady's web threads in a special tune that goes, "I'm not a gnat." More creepy are angler fish, who live at remote depths of the ocean. In most angler species, when the dinky male locates a female, he latches on. He latches with such conviction that his head more or less dissolves and he becomes a sperm-generating tumor with fins. (I can hear you wincing, ladies.) Sadder still is a species of deep-sea worm in which the eggish male, who is little more than a yolk-sucking sack of sperm, whiles away his entire life deep in the female's gut.

SCANTILY FURRED

At the very summit of the elongated animal before me is a shock of pale fur. There's quite a lot of it, sprouting from my skull like a lion's mane. It drapes down my back between my shoulder blades and comes to a stop where a groomer has snipped it. How long it would grow is anyone's guess. Each human's follicles are programmed to grow a hair for a set number of days before resting, then producing a new one that ejects the old. Both my sister and mother have allowed their fur to grow unsnipped, and their manes proved quite different. My sister's fades out at the shoulder while Mom's stretches halfway down her spine. Mom, whose follicles must be near round, grows the straightest mane; my sister's follicles are moderately flattened, producing curly fur; and my own follicles must be right in between.

That's what my fur looks like. But what is it for? Given the general nudity of the animal before me, half of my head is a surprising place to let loose with the fur making. It looks like an accent, like the mane that advertises a lion's masculinity, or the white tuft on a tamarin's head that helps fellow monkeys follow her gaze. Except that my fur serves neither purpose. It even impedes my view sometimes, an undesirable feature when tigers are stalking me, or tamarins are pointing their tufts at a jaguar.

The patchy pattern of human fur has yet to be explained by science. No one knows why long "terminal hairs" sprout mainly from my head, eyebrows, and the base of my limbs. Why am I not fully pelted, like a normal mammal? Stepping closer to the mirror, I'm reminded that I do, in fact, have fur all over. For one thing, my arms and legs sport a shimmer of terminal hair—sturdy, colored fur. And on closest inspection, tiny, transparent "vellus hairs" sprout from millions of follicles, head to toe. Vellus hairs are their own itty-bitty things, visible even on the silken underside of my wrist if I hold it correctly against the light. And they're fur. (I brace myself for the wrath of poodle owners everywhere when I assert that all fur is hair, all hair is fur, and even the quills on a porcupine and the scales on a pangolin are hair/fur. It's the same stuff.) All together, I have more fur follicles than a chimpanzee has. Furthermore, my pelage even attempts to keep me warm, retaining the piloerector muscles that hoist my hairs aloft when, for instance, I stand buck naked in a dim and chilly hallway in the name of science.

So perhaps the question isn't "Why no fur?" but rather "Why has the human pelt shrunken to something that wouldn't cover a mouse's modesty?" And that question reels us back to the dawn of humanity, when the hominids split from the other apes in our family tree. Six or seven million years ago, one ape species split into two species. One of those two new apes went on to diversify into today's great apes—the gorilla, the bonobo and chimpanzee, the orangutan—all of whom retain a respectable coat. The second evolved into the final great ape, me, the funnily furred. What event in proto-human history encouraged our kind to shrug off the body coiffeur?

And the answer is . . . Who knows? Our ancestral hominids left no written record of their lifestyle, and their fossilized skeletons are mum on the subject of how furlessness might have contributed to their survival. So when and why the fur shrank to this extravagant topknot remains a riddle. And that's the sort of knowledge vacuum that causes theorists to bubble with theories. Among them:

- We were so large bodied, and our environment so warm, that fur made us overheat.
- One of the sexes admired nakedness in the other, causing both genders to evolve toward nudity.
- Furlessness is a side effect of "neoteny," the human tendency to retain childlike features (like oversized heads and lifelong inquisitiveness).
- During an alleged aquatic phase in human evolution, body fur was, literally, a drag.
- We had to throw out the baby—our protective fur—to dispose of the bathwater—parasitic ticks and lice.

The first theory is the most popular. It relates to early hominids adapting to a new environment when they left the forest and took to the plains of Africa in pursuit of animal foods. If at first they galloped on four legs, then the fur on their backs protected them from the blazing sun. But eventually, to improve their hunting success, they evolved to scamper about on two legs. Over eons, they developed a profusion of sweat glands. (We inherited from these ancestors more sweat glands than any other critter on Earth—a few million.) Sweat cools an animal with tremendous efficiency, if it can evaporate quickly into the air. Because fur slows this sweat cooling, my ancestors gradually shed their fur, laying their damp skin bare.

This "sweat theory" suffers fewer soft spots than competing theories. For instance, if shedding the pelt foils parasites, why haven't more mammals done it? Or, if furlessness was meant to speed swimming, why have seals and otters retained fur? And why didn't early humans re-evolve fur when they rejoined the land mammals? The few bald mammals I can think of—elephants, rhinos, walruses, and whales among them—tend to be large, lending credence to the sweat theory. Because those species aren't endurance athletes, they sweat modestly, if at all. But the human, built for distance, sweats buckets. In sogginess per inch of hide, the human animal is bested only by the horse. And that's because the horse's northern range forces it to wear fur, which in turn necessitates heavier sweating. We tool-crazy hominids, by contrast, as we expanded into the horse's snowy latitudes, invented clothing, a layer of insulation that can be donned and doffed at will. Because we have tools, we don't need fur; and because we lack fur, we enjoy world-class cooling.

This explanation neglects the problematic hair on my head, though. The best theory here relates to my massive brain (which we'll get to soon enough). Theory goes: The head of my ancestral hominids, as they reared upright, was exposed to the tropical sun. As the brain of hominids evolved larger, this head fur grew in importance. A brain is a steaming wad of fat, a three-pound radiator. It's vulnerable to high temperatures and will fry out at 107.6°F (42°C). So it's insulated against the sun with a mop of fur, and it's cooled with a surfeit of sweat glands. That's the predominating theory, anyway: We shucked the body fur in order to cool our bodies better, but kept the head hair to prevent the brain from baking. My tresses shield my brain from the sun, the way a sheep's fleece keeps it cool in the desert. And the rest of my skin is open to the wind so I can sweat cool as I sprint after hamburgers on the plains of South Portland.

My mate, stripped down, is furrier than I am. Terminal hairs, powered by testosterone, sprout from his cheeks, jaw, neck, and chest, in addition to the tufts at the base of his limbs. Why males are furrier isn't clear. It could be that females have long preferred furry males, which resulted in furry ones outbreeding less-furry ones—a phenomenon known as a "sexual selection." The peacock's tail probably evolved thus: It has no function beyond looking fabulous. (The fact that many cultures currently abjure body fur on both females and males is immaterial—or not: If females were to abjure it for many generations, males might evolve to be as furless as females.)

Sexual selection would explain why fur patterns vary even from one region to another. As a rule, pale humans who evolved in Europe and the Middle East are hairiest, while south and eastern Asians, American Indians, and west Africans rank among the smoothest—

some males lack facial hair altogether. It may be that countless generations of Middle Eastern females preferred their males furry, while Asian females spurned the hirsute suitors. Also arguing for sexual selection is the fact that human fur erupts from our most odiferous body parts only as we reach breeding age. In humans, the apocrine glands concentrate at the base of the limbs and around the nipples on the chest, releasing fatty substances onto the hair shafts there. Although proof is slim, it's a reasonable thesis that the fur tufts on my body help to distribute my unique odor to potential mates. (What my mate can read in my odor, if he can detect it at all, is a question we'll tangle with in chapters 3 and 7.)

COLORATION OF SKIN AND FUR

If the distribution of my fur is a puzzle, its color is another. Mine resembles a field of dry grasses. "Applewood" was my grandmother's term, derived from the light and dark streaks in the tree. "Towhead" was my grandfather's term, a reference to the color of tow, aka flax, aka dead grass. Like my pale skin, pale fur is a radical departure from the prototype. Humans are thought to have evolved on the brightly lit savanna, where we adopted a dark hide under the dark fur that chimps and gorillas still boast. To this day, brown or black fur is the norm for humans in the latitudes where our species evolved. It's the groups that hiked *out* of Africa, and adjusted their tint to other climates, who need to explain their appearance. So, with no further ado, my wheaten locks drag me into the field of race.

Well, with a little ado: Many evolutionary biologists now assert there's no such thing as race, as different strains of humans. They argue there is more genetic variation between two northern Europeans, like me and my mate, than there is between me and my Mongolian friend, Bolortsetseg. This is, to my mind, a mathematical gimmick. Certainly, Bolor and I share oodles of genes that regulate our kidneys and our fingernails. Those fundamental features aren't subject to the climate pressures that drive evolution. But most humans, including me, can effortlessly pick out where Bolor's and my genomes differ, too. We can look at ourselves and see something that's eluding the theorists: Whatever differences do exist in our DNA, they're having a big impact on the way we look. They're affecting not just skin colors, but entire suites of characteristics, involving eyelids, noses, lips, skin, hair, and even body shape and size. And if we're like every other animal on Earth, we came by many of these differences honestly: Groups of isolated humans evolved in distinct environments, which rewarded distinct arrangements of their DNA. So, while one scientist has characterized racial variation in the genome as being "scientifically and mathematically trivial," these differences are certainly not ecologically trivial. The relationship of a human body to the habitat in which it evolved—or in which many generations of its ancestors evolved—is far too marvelous to ignore. Besides, we humans view other animals in exactly this way: as races, different in color, shape, size, and geography, but grouped under one species. For instance, the black bear in Florida weighs three hundred pounds and eats lots of vegetation, while the black bear in Newfoundland weighs four hundred pounds and eats moose and caribou, and the black bear in Alaska is often brown.

Now, back to my distinct shape and color, which I insist do exist. In the mirror, my skin is pale, which unkindly accentuates every bump and blemish. This is typical of the northern European package. My skin is pale, my fur is pale, my eyes are pale, my nose is long, my cheekbones narrow, and my lips thin. I'm close to the palest end of the human spectrum. But the European human is unusual in that it comes with a lot of color options. Within my immediate family the skin is reliably pale, but I can find eyes in brown, green, and hazel, and fur that's reddish, brown, and nearly black. This grab bag may be related to the varying levels of melanin in the northern animals, or it could relate to a few freakish but harmless mutations. Pretty much all the other groups of humans on Earth have retained brown eyes and dark fur.

So why is my face so wan compared to humans from Africa, India, and Australia? Considering how long the subject of skin color has transfixed humanity, it has taken a very long time for a solid theory about it to emerge. Until recently, the best guess was that dark skin dominated at the equator to fight skin cancer, and pale skin evolved to the north and south to hasten the body's manufacture of vitamin D from scarce sunlight. But skin cancer is a red herring. It takes so long to develop that most victims don't perish until after they've raised offspring. It should have no effect on evolution.

Vitamin B, or folate, is a different story. It has inspired Nina Jablonski, a pale, American paleoanthropologist, to pen a robust proposal. She and her coresearcher-mate began with one of the human animal's more advanced tools, an orbiting satellite that looks down to measure the amount of ultraviolet (UV) radiation reaching the earth. Equipped with a map of UV intensities, they next set out to see what shade of human skin has evolved under the brightest spots and the dimmest. Then onto that map they plotted the color of human hide. And yes, human skin is darkest where UV is the strongest. But skin cancer is not the reason, Jablonski says. Folate is.

The human acquires folate through the diet. But when UV radiation penetrates the skin, it breaks that folate apart. Human females with folate deficiency are prone to miscarriage, and their surviving offspring can suffer fatal nerve damage. Males deficient in folate produce lousy sperm. These are the sort of effects that steer evolution. As hominids evolved under the African sun, those individuals who chanced to have more melanin in their skin were protected from UV and maintained healthier levels of folate. They produced more offspring, and thus won more chances to send their genes forward in the human race.

So, if the need to protect folate rewards dark skin, then what caused melanin to wane in humans like my ancestors, who migrated to gloomy spots on the UV map? Well, the human animal has a conflicted relationship with sunlight. As much as we need to keep UV out, we also need it to penetrate the skin for the manufacture of vitamin D. Vitamin D is crucial to the construction of bones and to healthy blood chemistry. We must have it, and the sun is the source.

So, Jablonski concluded, human skin color strikes a fine balance. It must be dark enough to preserve folate, yet pale enough to produce vitamin D. It's a tidy theory: In sunny places, humans can afford dark skin because they'll still get their vitamin D. In dim spots, the threat to folate diminishes, so skin can afford to lighten. It even explains why the skin of females is generally a shade lighter than males': Females need more vitamin D to build offspring and nourish them with milk. (I suppose it might also explain the dearth of female facial hair: If I had a beard, it would reduce my D-building capacity.)

The UV map doesn't predict human skin color perfectly. Occasionally, Jablonski found light-skinned humans in high-UV zones and dark-skinned humans in dim latitudes. Some mismatches result because our restless species has wandered to a new habitat so recently that we haven't yet evolved the ideal coloration, Jablonski says. (English migrants to Australia, for instance, have yet to adapt to the intense UV there. They presumably enjoy indestructible bones, but they also suffer terribly from melanoma.) In other cases, such as the dark-skinned Inuit, the diet may be so rich in vitamin D (from fish) that skin can afford to stay darker, safeguarding folate. So my ghostly skin is a certificate of authenticity: I'm adapted to the cloudy climes.

If I accept that humans come in distinct varieties—and I do—then what shall I call them? Are my Mongolian friend and I two different "subspecies" of *Homo sapiens*? That's the term biologists use to designate groups of animals that look a bit different but easily interbreed. For instance, the red-tailed hawk, *Buteo jamaicensis*, which is native to the Caribbean and Americas, splits into sixteen subspecies, including *B. j. calurus*, whose head is solid brown, *B. j. kriderii*, who's pale all over, and *B. j. harlani*, with a mottled breast. None of these is more or less a red-tailed hawk than another. They're just, you know, different.

Or perhaps Bolortsetseg and I are two "morphs." When individuals within a species differ mainly in color, biologists use this term. Many hawk species come in two color morphs, light and dark. The North American gray squirrel sometimes turns out black. A jewel-toned Australian finch is available in three color morphs: with ruby face, orange face, or black face. Domesticated animals are more flamboyant in their morphic variety. For instance: We've bred dogs to vary from the size of a pony to the size of a breakfast pastry. We've created domestic rabbits in every natural hue, and with ear styles that wouldn't pass the laugh test in the wild.

So what word suits? "Morph" seems too small to cover the many differences between Bolortsetseg and me. Our coloration differs strongly, but so do more structural elements, like our cheekbones, noses, lips, and eyelids. But the word "subspecies" suffers from the fact that *Homo sapiens* is a status-conscious animal, and none of us wants to be a sub-anything. I'll try to make peace with the word "race," in the fluid spirit of UK biologist Armand Marie Leroi. He refers to the concept of race as "a shorthand that enables us to speak sensibly, though with no great precision, about genetic rather than cultural or political differences."

In recent centuries, humans have typically discriminated among three races of modern *Homo sapiens:* Asian, Black, and Caucasian. But most humans can spot finer divisions within these ABCs. To my eye, my own British Isles features look more lumpy than the angular faces of the French, but a bit less lumpy than the features of pale-skinned Russians. Likewise, to my eye, Australian Aborigines, with strong brow ridges and wide noses, look completely different from heart-faced Bushmen of southern Africa, who in turn look totally different from the diminutive Baka Pygmies of Cameroon. It's this finely divided concept of race that Leroi champions. There are, he argues, as many races as you need to parse out in order to discuss human variation.

It works for me. Under this paradigm I can call myself a member of, from smallest subset to largest, the British Isles race, and the northern European race, the Caucasoid race, and the human race. It's imprecise, fluid, and maybe even politically incorrect. But it's an honest reflection of how my skin, fur, and facial bones took shape in an isolated pool of humans breeding in a cloudy, cool environment.

The subject of human diversity churns with emotion, but plenty of other animals have also evolved distinct races in distinct environments. The races of red-tailed hawk vary not just in color, but also in size and breeding habits, depending on the pressures of the environment. The tiny leopard cat, a mottled beauty who roams much of Asia, comes in eleven subspecies, with camouflage fur that runs from gold in the south to gray in the north, and whose size in the coldest part of its range is one-third larger than the southern races. The green-backed heron, whose range circles the tropics, is available in a bewildering twenty to thirty races. Whenever a population is isolated, it will continue to evolve, drifting slowly out of similarity with the other populations of its species. My low-melanin face is proof.

I'm not making great speed down the body in the mirror, but I suppose that in a hominid the action concentrates at the top. I've ruffled my fur and pondered my pallor. The next obvious feature is, well, my features. This cluster of apertures does so much work that it's best to skip past them for now and come back to them in chapter 3. For now, suffice it to say it's no accident that four of my five major sensory organs are crammed together in my head. It's human habit to think of the ears, eyes, nose, and mouth as "the face." But taken separately, each is a sensor: My eye is a moist camera, my nose protrudes to sample the air, my ears funnel vibrations into my head. That face in the mirror is really a cluster of tools gathered around a motherboard, a CPU, the brain. And in the human, the brain, too, deserves a chapter all its own. In terms of the vision before me in the mirror, the next specialty of my species is the trait that puts the "mamm" in mammal.

PROMINENT MAMMARY GLANDS

Departing from the crowning glory, we slide down the neck (I'll revisit that feature in conjunction with the legs and posture). And we quickly run aground on a blatant discrepancy between males and females—a pair of blatant discrepancies, really. Were my mate here beside me, we would see his two nipples sitting flat on his pectoral muscles, useless as an eye on the bottom of a foot. Formed early in the life of a fetus, nipples appear on most male mammals. (This may be a fillip of evolution, a mutation whose cost is so petty that a nipple-free male would have no advantage over a nippled one, and hence could make no evolutionary headway.) But here on my female chest, each nipple sits atop a hemisphere of fat. These fat stores wobble when I walk, joggle when I jog, and generally call attention to themselves. They, along with my waist-to-hip ratio, can telegraph my gender to an observer a hundred yards away.

Whether that's why my enlarged glands evolved is open to debate. The human female's mammary glands are, like the human fur pattern, another oddity of the animal kingdom. In most mammals, the milk glands swell only for as long as the mother is nursing young. When the offspring are weaned, the glands deflate and get themselves out of the way. Mine, which materialized along with my other secondary sex characteristics (tufts of fur, a wider pelvis, fat pads on the thighs and buttocks), are always puffed with fat. I see no advantage to this arrangement. I could easily store this fat in a thin layer under the skin or on my thighs. And, therefore, it is quite likely that the whimsical force of sexual selection has been at work. My puffy breasts may be announcing to the world's males that I have garnered sufficient fat to support hearty offspring. Evolution would reward this architecture if my in-your-face advertising attracted a better selection of males. The ability of larger breasts to store more milk might also drive evolution: Large-breasted females can stretch the time between feedings of an infant; that, in turn, reduces the frequency of the infant's distress calls, which reduces the chance of a predator hearing and locating the infant. In that case my lineage might be in trouble. From the look of things in the mirror, my offspring would issue frequent distress calls. But perhaps the trade-off is that I, relatively unencumbered, would be able to pick up the young squawker and run like the wind, e attracted any predators. The *least* favorite theory I come across longs to dear Desmond Morris, who in his classic *Naked Ape* proposes that breasts evolved to mimic buttocks, thereby to inspire a male's mating urges from both front and back. It's my humble opinion that mammary lust is a cultural phenomenon. It's my humble observation that in cultures where clothing is minimal—the Yanomami of the Amazon, the Himba of Namibia, the French on the Riviera—it's really only the mating parts that humans specially strive to obscure.

But in my culture most females keep the mammary glands under wraps. I'm getting self-conscious. Let's exit my cleavage, what there is of it, and descend to the abdomen.

DISTRIBUTION OF STORED ENERGY . . . OKAY, FAT

Sliding downward from my narrowed female waist, we arrive (sooner than I'd prefer) at what a physiologist would call my "primary energy-storage tissues." It is not an accident that my hands hang so as to block my (your—everyone's) view of these tissues. Within my culture, these tissues are mortifying. My animosity for body fat is a recent development in human history. But we'll examine that in chapter 6. Here, I shall try to cultivate an appreciation for lipids—not just for how they serve the human animal, but also for why they agglomerate where they do.

Fat is not, after all, gratuitous. The glorious human brain can't function without cholesterol, a special fat cooked up by brain cells and the liver. Throughout the body, additional cholesterol insulates the nerve fibers and builds cell membranes. It's also key to the manufacture of the sex hormones I wouldn't want to live without. And that's but one form of fat.

More prosaic is this layer that wraps my hips. This stuff is just a mass of specialized cells, called adipocytes, each holding a droplet of oil. When I eat more food than my body can use, it's converted to oil and stowed in an adipocyte. Evidently I've done this a few times.

Then when I eat less than I need, oil is pulled from storage, and burned. This stored fat of mine is like a bank account. And for all mammals, there's a minimum below which the balance cannot fall without incurring a penalty. Most members of the warm-blooded crowd need between 4 and 8 percent body fat on hand to support normal activity and reproduction. I am probably carrying 30 percent, because . . . I need to have a cookie and think about that.

One reason I have so much body fat is that I'm female. A fit human male is about 15 percent fat, while a female, on whom offspring depend for nourishment, is 23 percent. I carry even more than that because . . . because cookies are too damned easy to capture.

Anyway, many humans no longer need to bank much more than the minimum of oil. Increasingly, we use tools to store food *outside* the body. This mimics one of nature's classic strategies, whereby squirrels, mice, rats, and many other small critters stash food for later use. Most humans now use tools to preserve food, so that we needn't forage for it in the wild. I might be a bit slimmer if humans had adopted the migratory method instead of the food-storage method. Animals like caribou, wildebeest, many birds, and even whales have evolved to migrate around the globe in pursuit of a shifting food supply. As food fades in one part of the world, they travel to where it's flourishing.

Alas, although we've learned to cache food, humans still behave like one of nature's true fatties. A few animals have adapted to pack a prodigious lunch under the skin, then live off it for months. The camel, some bears, the hedgehog, the badger, and migratory birds can become naturally obese. What these pudgers all have in common is a sporadic food supply. They make hay while the sun shines, and they carry that hay with them through the stormy weeks and months to come. They're as adept at fasting as they are at feasting. Although humans have proven to be prodigious feasters, the fasting part we just can't master.

If I had the lifestyle of a polar bear, my excess stored energy wouldn't be a problem. During the winter sealing season, these animals accumulate a spectacular burden of fat. The sea ice melts in spring, forcing the bears onto terra firma, where they enter a dopey "walking hibernation" to save energy. The fattest of the fat are the breeding females. A female who's pregnant at this point faces a nine-month fast during which she'll gestate in a den, then give birth and nurse cubs. In preparation, she will pile on four hundred to eight hundred pounds of fat, and nearly become incapacitated. A pregnant female looks like a snowball.

I am not one of nature's natural fatties. Because a human moves around more than a sleepy polar bear, the weight of extra fat taxes human joints. Because my body chemistry can't adjust to excess fat, obesity threatens humans, but not polar bears, with cancer, diabetes, and heart disease. Unfortunately, the mere fact that I'm not designed for fatness doesn't prevent me from swelling. I came into this world with a huge number of small, empty adipocytes distributed under my skin, between my muscles, and around my organs. They yearn to be full. I yearn for them to be empty. You could cut the tension with a butter knife.

It's not just me. Fat "generally behaves in an unpredictable and inconvenient way," writes fat scholar Caroline Pond, a lecturer at the Open University in the United Kingdom, in her book, *The Fats of Life*. From one human to the next, fat cells congregate in different patterns. The distribution of my energy stores is partly genetic. Among the females in my family, it's common to store energy in the hips and thighs, and not in the mammary glands. But fat's shape has a nongenetic component as well, leaving plenty of room for individual expression. We enjoy, in fact, more expressive fat than any other species. In other animals, fat behaves more predictably and conveniently.

My fat does obey a few rules. The most important one is that it must stay out of the way. I can't store it around my neck, for instance, as that would hamper my ability to look down. It can't swaddle my fingers, whose thin tips do delicate tasks. It mustn't coat my head, lest my brain overheat. It mustn't thicken the inner thigh, or my ability to chase prey will suffer. Given all these restrictions, one safe place for both males and females to stow fat is in a smooth layer, just under the skin.

Subcutaneous fat isn't standard for mammals. We expect to see that layer on marine mammals like seals, orcas, and whales. But most others direct their fat to designated depots. The camel and the bison concentrate theirs in the hump; the beaver and the platypus fill their tails. The gray squirrel keeps half his total fat deep in the gut, and birds stack lots of it at the wishbone, perhaps to keep fuel handy for the wing muscles. Primates like chimps and gorillas favor fattening the paunch. Many other mammals lay in a pad of fat over their shoulders. Although mine congregates on the hips, it also pads the entire underside of my hide.

The anomaly of my subcutaneous fat helped inspire one of the weirdest theories in human evolution. Try as I might, I can't describe the aquatic ape theory with a straight face. Here it is in a nutshell: Somewhere back in the early millions of our humanhood, we took to the sea like so many otters. We caught fish with our . . . fingernails? (We had long since downsized the canine teeth that many predators rely on for fighting and feeding.) As this theory would have it, we lost our body fur because it retarded our swimming; for insulation, we developed a dolphinny layer of fat under the skin. We waded or swam after our prey, with only our furry heads protruding into the sunlight.

Now, I love dolphins as much as the next *Homo*, but this just doesn't wash. For one thing, the fat layer on a human is most un-dolphinny. Real aquatic mammals like seals and whales are born with a sleek layer of fat, and they maintain it as they grow. (And it's not for insulation, Caroline Pond suspects. Because marine-mammal fat is not distributed in a uniform layer, she guesses it has more to do with streamlining and fueling nearby muscles.) Human fat, by contrast, waltzes around the body during the course of a life span. I was a chubby baby, with fuel packed on to warm me and smooth my transition to nursing, which can take a few days to grasp. By age nine I was outgrowing my fat supplies so fast that I resembled a bicycle frame minus the padded seat. At eighteen, I began laying in fuel against those fat-draining tasks of pregnancy and nursing. And one of these years, my shifting hormones will move fat to my waist. I assure you, the fat topography of my mature body does absolutely nothing for my hydrodynamics.

There are other obstacles to my aquatic success, as well. My flat snout, even if it were fitted with respectable canine teeth, would expose my nose and eyes to thrashing prey when I hunted—unless I was a tool-wielding aquatic ape. And then there are my fingers, which aquaticists claim are webbed. In reality they do their best swimming when the alleged webs are tucked away between my cupped fingers. Regardless, the aquatic ape theory continues to cavort in the sea of ideas, apparently facing no imminent risk of extinction.

Back to my fat. In addition to the subcutaneous stuff, I bear some conspicuous bumps fore and aft. My mate's fat gels elsewhere. His, what there is of it, pads the abdomen, as it does in other primates. Had he more, it might accrue on his shoulders and at the back of his neck. Caroline Pond speculates that females can't deposit their healthy bank account of fat on the abdomen, as males do, because it might overheat a baking fetus. On the rump my fat is, if not out of mind, at least out of my way.

Pond also believes that my mammary-fat storage is not just sex-specific, but also sexy. Her theory flies in the face of the feminist argument that mammary glands are feeding appendages and not sex toys. The argument is intriguing: Just 4 percent of a healthy female's body fat lodges in the mammary glands, on average. Yet this 4 percent makes a big visual bang for the buck. And that, Pond says, is a dead giveaway for a sexually selected feature: Like the nodding feathers on the head of a California quail or the dark mane of a lion, my bobbling mammary glands are more flashy than they are practical. Perhaps they signal my readiness to nurse young.

The final and most unfair difference between male and female fat storage regards the distribution of collagen. This pains me, but I'll just come out and say it: The energy-storage tissue on my leg is dimpled. Probably for most of human history, and definitely in

Renaissance Europe, this was considered a great asset. That was then. Biologically speaking, the dimples come from the female arrangement of collagen, the fibers that anchor skin to the underlying muscle, running through the fat in between. In a female leg like mine, collagen fibers pass through the fat in a straight line, like nails holding a mattress to a wall. But there are so few nails holding up my fat that the mattress sags in between them. Adding insult to injury, my female estrogen works on collagen over time, contracting the fibers. This exacerbates the bulging. In males, by contrast, the collagen nails crisscross one another, giving greater stability and smoothness to the mattress. Why did evolution lead me to this bumpy state? No clue. I suppose that thigh dimples, like breasts, could signal that a female has stored sufficient fat to grow a dimpled baby. Stranger traits, if not less fashionable ones, have evolved.

If there's any justice in this fat situation, it's in the way fat moves around as humans age. The majority of males will actually develop breasts of their own as their hormones drift toward a more female mix in old age. Of course, my own fat will migrate, too, deflating my hips and padding my waist out of existence. With time, the distant silhouettes of males and females come to telegraph the same message: past prime.

THE FORELIMBS, PRIMARY MEANS OF DEFENSE

Speaking of my hips, that hand poised to obscure my rump deserves credit for more than obfuscation. Why, that hand is also my state-of-the-art weapons system—five sticks of thin skin and fragile bone, each tipped with a wafer of wannabe claw. I'm trying to count the times I've used it successfully. I guess I balled the hand into a club and knocked my brother down once. And I remember whacking a boyfriend who was breaking up with me in a spectacularly stupid way. In both cases, my assault was mainly intended to punctuate a barrage of verbiage I had wearied of. Which is about all my little sticks are good for. Until you give the human animal tools, it is pitifully armed, and not dangerous.

This isn't normal, among primates. Cousin Chimp's weapon of choice is his large teeth. A chimp attack on a human male in 2005 put this animal's armaments on display. The human, a visitor to a private chimp sanctuary in California, was bitten almost to death. Two male chimps chewed off most of his fingers, his gonads, and much of his face, and they bit a foot clean off his body. The human survived, but probably only because another human attacked the two chimps with a deadly tool. Gorillas and orangutans, as well as the other primates, also rely primarily on their teeth to wound and kill their rivals.

A fighting ape or monkey will bite any bit of anatomy that presents itself. But he will make an extra effort to savage toes and fingers. And primates seem driven to damage one other appendage, as well. Time and again, biologists have noticed that scrapping males, if they get a chance, will slash at an enemy's testicles. I suppose it's a sound evolutionary strategy. If you can't kill a rival, you might at least eliminate him from the pool of males competing with you for breeding opportunities.

I don't believe my jaws are up to the challenge of biting off human fingers, let alone severing an ankle. Humans don't fight much with our teeth. Generally, we prefer metal tools, and generally the longer the reach of the tool, the better. When a human does tangle without tools, the weapon of choice is the hands. A male boxer in top condition can punch with a force of a thirteen-pound mallet swung at twenty miles an hour. Even an amateur human attacking with the hand can break a rival's mandible—in fact, the male fist is the most common cause of broken jaws. Less often, an attacker can strike the skull hard enough to break bone or cause an internal brain injury. Any of these attacks could conceivably lead to death through suffocation, infection, blindness, or starvation, under suitably primitive conditions. But all things considered, it's pretty hard for a human to kill a rival with his bare hands. In my culture, only one killing in fourteen is accomplished without tools.

My teeth are my secondary weapon, and they can be more damaging than my claws, in a sort of backhanded way. To explain, let me abscond for a moment to eastern Indonesia, home of the Komodo dragon. While this 350-pound lizard prefers to kill prey outright, if a deer or large boar can shake the dragon's bite, he doesn't fret. While the prey animal counts its lucky stars and scoots away, in the bite wound special bacteria fester. A day or two later, the animal drops dead. Then the Komodo dragon, who has followed, eats.

My species has a similar tactic in its arsenal. "Fight bite" is a condition that results when a human fist punches a human tooth. You might think of it as "Komodo knuckle." At first the attacker appears to have gotten the upper hand. But follow him around for a few days. When the punchee's tooth split the skin on the attacker's knuckle, his spit entered the wound. When the fist unclenched, the tendon retracted, pulling spit and bacteria into the joint. From the outside, the wound appeared minor, so no one thought much about it. And 70 to 90 percent of the time, such a wound heals. But when it doesn't, the infection is among the most foul. Surgeons must carve through hot, red flesh to open the joint and flush out pus. Sometimes they must do this over and over. The bones may become infected. In parts of the world where surgeons are in short supply, one in five fight-bite cases ends with amputation. Is fight bite an evolved weapon, like Komodo-dragon bite? We may have evolved to get the last laugh, or our toxic teeth may be a lucky break. Either way, like our frail hands and claws, the mouth is a weak weapon and a strong argument for the early adoption of tools.

HIND LIMBS: DEVISED FOR ENDURANCE

Leaving my weapons, I reach the knees. Oh, my brave and battered knees. They're chipped with scars from a childhood outdoors. The skin is thickened because, despite the human's two-legged reputation, we still do scuttle about on all fours a fair amount. I kneel to forage for vegetables in the garden or in the fridge, to stoke the woodstove, to scrub the floors of the shelter. Out of sight under the skin and scars, the patella is misaligned, causing sporadic agony. This human ran too many miles in poor shoes on hard roads, and now must treat the knees as though they were made of glass. But from a distance, my knees look perfectly—well, they look ridiculous.

Knees are comic on any animal. On the elephant, they're the stiff-looking joint that points forward. In a chicken or a stork, they're the meeting of the thigh and drumstick, folded high against the ribs. On a rabbit, the knee is at the front edge of the big, round hopper. Bees' knees are rather conceptual. And on me, the knee is a bulge where the femur balances right atop the head of the tibia. No other mammal's leg looks like mine. In skeletal form, it appears horribly uncomfortable, as though the knee would be pounded to paste in a couple of miles. But the entire human body has been remodeled to make the most of this joint. From head to toe, I'm built to run, with the knee forming a hinge that allows my leg to swing forward for the next stride.

The running ape hypothesis is new but impressively complete, one of those ideas that makes everyone feel a little stupid for not thinking of it earlier. The two American fellows who fleshed it out, a paleontologist and a biomechanics researcher, have compiled a list of twenty-six physical features that contribute to my running prowess. And here I must be clear about my choice of pronouns. By "my prowess," I mean something like "the prowess of my species," or "my potential for prowess." My personal prowess is wholly theoretical. Too much running at an early age, in the slivers of rubber that once passed for running shoes, on the slanted asphalt of back roads, has put my distance-running days far behind me. But theoretically, if my hip joints and knees and oft-sprained ankles weren't ruined, I could run a long way.

Obviously running isn't brain surgery. Cheetahs and pronghorn antelopes can hit seventy miles an hour, after all. The fastest humans can go only one-third that fast. But those are sprinting speeds, for cheetah, pronghorn, and human alike. My strength lies in my ability to run an hour, or two, or even three, mile after mile. And that is a rare ability. Only a few animals enjoy membership in the distance-runners' club. They are the horse, wolves (and domestic dogs), African hunting dogs, the hyena, the migrating wildebeest, and we *Homo sapiens.*

How do we do it? Differently from the other marathoning species, that's for sure. We do share one of our twenty-six features with the other creatures, but each species evolved this trait individually, in a process called "convergent evolution." At the base of my skull I can feel a knob. Just below that are attached the massive rubber bands of the nuchal ligament. The bands, along with a modest fan of muscle, anchor my head to my back, and prevent my skull from bouncing when I run. I can keep my eyes on the prize even as my body is bounding over hill and dale. Pigs, among a thousand other animals, cannot. I remember the pigs of my childhood would break into a trot when I approached with their breakfast: As each stiff leg hammered down, the cone of the head bounced and yawed, the ears flapping and the eyes rolling to keep track of their meal.

My neck is another of the twenty-six running features. A chimpanzee's neck is bound tightly against the shoulders by muscle; her waist is rigid; her pelvis is wide; and her forearms are long. A running chimp is as graceful as a tumbling brick. But my human neck is long and untethered from my shoulders. When I swing a leg forward, I can counterbalance by swinging the opposite shoulder and arm forward. My head, on its flexible column, rises above the tumult. Down below, my elongated waist allows the torso to swivel over striding legs, and my narrow pelvis permits my weight to shift from one leg to the other without waddling. To the rear, my gluteus maximus, one of the body's biggest muscles, pulls my torso erect over the legs. Running, one running scholar has said, is a controlled fall, and it's the glutes that prevent the fall from following through. Cousin Chimp, I've noticed, has a skinny butt worth nothing at all.

Below the butt, a second great rubber band toils away at the back of my leg. The Achilles tendon connects my calf muscle to the heel. It doesn't do much for the chimpanzees, who have a shorter version; nor does it help me much when I'm walking. But when I break into a run, this and the other tendons in my legs and feet act as springs. They alternately store and release energy, recycling 50 percent of the effort I expend. Like the nuchal ligament, this reduces the burden on muscle and stretches my endurance. The bottom of my femur and the top of my tibia flare wide where they meet, to spread the force of impact.

My feet are unprecedented in nature, each a mesh of twenty-six bones and ligaments built to absorb the full weight of a human running for miles on end. My ankles are wrapped in tendons and muscle to stabilize the bones, although I've lost count of the times I've sprained and strained them anyway. The left one is still bulky from when I fell a year ago and ripped some of those tendons loose from the bone. Further down, my spring-loaded plantar arches, unique to the human, flatten each time my foot hits the ground, then contract to bounce me forward. My own arches are high and strong, thanks in part to a barefoot childhood that strengthened them, and in part to my healthy body weight. (Arches form in childhood and can be squashed by stiff shoes or a heavy body.) Out at the end, my toes are short and stiff. One cost of running on two legs is that I've lost the grasping toe that other primates use to climb. I need mine to make the minute corrections to my balance that keep a large body from toppling off its two legs.

And the list goes on. The zillions of sweat glands distributed over my long body keep me cool as the miles go by. Even my short snout is credited with balancing my head more squarely over the spine.

It all adds up to a glossy picture of Running Ape, an animal evolved to go an enormous distance without rest. And when biologists compare humans to other distance runners, we travel at the front of the pack. The best human runners can outlast a good horse. And even the wildebeest, workhorse of the migratory mammals, flags sooner than the horse, so I suppose we beat him, too. On a day-to-day basis, many human runners cover the same distance—six to twelve miles—as do African hunting dogs, wolves, and hyenas. I find this remarkable. Why on Earth should we have evolved to be the elite long-haul runners?

The classic theory is that running evolved as a by-product of walking upright. Running achieves the same goal, only faster. But why distance running, this unusual ability to run for hours? One idea is that when protohumans evolved on the plains of Africa they deciphered the language of vultures. By walking to the spot beneath the circling birds, they could score a rich meal of meat, courtesy of some lionlike predator. And by running to the spot, they could beat the other scavengers to the punch. A competing theory suggests ancient hominids ran for the same reason the wolf runs: to catch up with live prey. The Bushmen of the Kalahari still hunt antelope that way. For hours they'll chase the animal, preventing it from resting or cooling off. Antelope, and most other animals, will tire before we will.

Well, actually, if they come up against *me*, they'll be just fine. That's because evolution isn't finished with me yet. My lower back aches. My knees are corroded. My hip joints began flaring into pain when I was sixteen, and they continue to protest to this day. All these ills relate to the way the human skeleton has been reeled up to a standing position. I'm not an isolated whiner. Eighty percent of my countrymen will suffer back pain in their lifetime. Half to three-quarters of all humans on Earth will probably endure degraded knee joints. The hip joint, too, frequently degenerates into arthritic wretchedness as the body ages. All these joint problems vary between groups of human. In some cases, genes seem responsible—hand and hip arthritis seem to run in families. In other cases, cultural habits can press heavily on a human body: A study of Chinese and U.S. humans found that arthritis of the knee is much more common in China, where it is traditional to squat while resting—a position that stresses the joint. It seems that the human version of two-legged locomotion is a work in progress.

But all animals degenerate as they age. Whenever skeletal joints are imperfectly aligned or poorly padded, arthritis will arise, building telltale spurs of bone. Dinosaur fossils show it. Chimps in Africa endure it. Wolves, and my dog who tore a ligament in his leg, get it. Even a whale who recently died in the Thames River in England had dreadful arthritis of the neck and skull when he was autopsied. One would think a watery lifestyle might spare a creature's joints. But perhaps the whales are still haunted by their prehistory as land-dwelling mammals, and their skeletons haven't fully adapted to floating.

We have come from head to toe. And here, at the very tip of my anatomy, is another key for differentiating a human from the other great apes: We can't leave our bodies alone.

SELF-MODIFICATION

This subject first struck me in a Costa Rican museum where I encountered small cylinders of carved clay. Perhaps three inches long, they were used by prehistoric humans to roll a design in pigment across the skin. Left behind would have been a colored pattern. For a moment I thought: How thoroughly delightful and primitive! They decorated their bodies!

Then I noticed my own reflection in the glass. I took a sober survey. I was delightful and primitive from top to toe. I had paid someone to sculpt my head fur. I wore a metal chain around my neck, and from that hung my "traveling stone," a bit of hometown geology about which I held certain superstitions. My ears also bore metal, which pierced the flesh. And two of my fingers were circled with additional rings of metal and stone. I had circumscribbled my eyes with dark pigment and tinted my lips red. I had even overwashed my natural odor with a distillation of flowers and spices. My toenails were pink.

Humans so frequently alter their bodies that a physical description of one must incorporate decoration. You would be hard-pressed to spot a human in the field who is unadorned by toe paint, earlobe stretchers, metal hoops at the finger, neck, arm, or leg, a bone through the nose, a gourd on the penis, bleach in the hair, dung in the hair, studs in the tongue, henna on the fingers, feathers on the ankles, or "Mom" on the biceps.

Why, I wondered, gazing at those clay rollers in Costa Rica, does my species see itself as a blank canvas, begging for a coat of paint, a hoop of metal, a filed tooth, a fan of feathers?

I know why *I* do it. It's partly superstition, a belief that certain forms of metal and stone prevent ill fortune. It's also partly mating related: My face painting began as I reached reproductive age, which is also when I had holes punched in my earlobes. In many cultures besides mine, youth make a radical change to their appearance as they reach reproductive age. A female may cut her hair short for the first time or have her skin slit to form ritual scars. She may begin to blacken or file her teeth, or, as I've read of some tribes in the central highlands of Vietnam, she may take up a rock and chip out her front teeth altogether. A pubertal boy might slice his lower lip to accommodate an ivory plug, tattoo his face, or fillet the underside of his penis, widening it and permanently altering the direction in which he urinates.

Body alterations also telegraph status. Although my own culture isn't easy to read this way, others make it simple: Many brass rings stretching a female's neck? Excellent family. Feet bound from girlhood to resemble croissants? Family so rich she needn't work. A male's face completely tattooed with spirals? Important guy.

And as they do the world around, my cosmetics also advertise my tribal affiliation. Female fingernails are tremendously eloquent in my culture. One female can learn much from a glance at another female's nails: Are they French manicured, filed straight or to a curve, long and natural, long and acrylic, or ornamented with flowers, sparkles, or American flags? Each style, including my own (ragged and paint-spattered at the moment), aligns the female with a different tribe.

And the playful breezes of fashion certainly play a part. Human fur is especially susceptible to whimsy, and it accounts for some of the weirdest modifications ever made to the human animal. Ancient Egyptians were lukewarm on the stuff, and many who could afford to shaved off every strand, replacing it with wigs. Elizabethan English females of means tore out only the front inch or so to elevate the brow; or cut it all off and collected wigs. Cultures lacking metal scissors have been known to barber with shark teeth, shards of obsidian stone, or sharp-edged reeds. My own identity is so tangled up with my fur that I've altered its very color. Once a bright blond, my fur has dulled with the years. And I just couldn't stand the sensation of "losing my looks." So now I pay a fellow human to streak my head with toxic chemicals, stripping out some of the brown. It's a curious phenomenon, this hair coloring and uncoloring. The Dinka of Sudan lighten their black fur with cow urine. In India many aging males disguise their graying fur with orange plant pigments. Fur style is serious business for a human.

Of course, body modification is partly about pleasing my own eye, too. My earliest recollection of the "canvas potential" of my body came after scratching a leg as a child. Under the scratch my Caucasian skin turned first pink, then blanched to white. Cool! Using my fingernail, I made a new line, this one serpentine. Cool! Next, I made two lines, side by side. . . . But whether these are the same urges that first moved hominids to alter their bodies, I can't say.

Anthropologists tend to think it starts with self-awareness. In other words, a bear is so ignorant of its own self that it would never think to alter that self. Even the decorator crab, who transplants colorful reef flora onto his carapace, does so instinctively, because evolution has rewarded those crabs whose odd behavior camouflaged them from predators. Glimmers of self-awareness do twinkle in apes. A chimpanzee presented with a mirror does recognize herself. She may even put leaves on her head or otherwise toy with her appearance.

She may turn to check out the parts she's never had a goo
One might also argue that a gorilla using sign language to
her emotions displays self-awareness. And recently, a few
passed the mirror test: Instead of reacting as though th
stranger, they used their reflection as an opportunity to explore their bodies. Dolphins and magpies, too, have shown an inkling of self-knowledge. Beyond these animals, I know of no other who displays an appreciation of itself. However it was that we humans were first inspired to self-alter—to attract mates, or to advertise tribal membership or high status, or through toying with our sense of self—we're now irrepressible. All surfaces are fair game, and often the alterations are for keeps.

The only permanent changes I've made are the small holes in my earlobes. But consider the male animal who shares my shelter. If we direct our attention to his genital region, we'll find . . . something missing. No one knows when humans began removing part of the penis. Egyptians were apparently doing it six thousand years ago, but it could have started long before that. Many humans still conduct circumcision on male offspring, citing a list of beliefs that range from social to religious. New research offers a more biological explanation: Circumcised males are less vulnerable to infection with HIV. The virus apparently prefers to enter through the frail cells of the foreskin. If circumcision does indeed reduce the risk of males contracting fatal diseases, that could well have kicked "cultural evolution" into gear long, long ago: Those groups of humans who practiced the cultural behavior would enjoy better survival rates.

Equally obscure is the origin of tattooing, the injection of pigment under the skin. Tattooing is practiced around the world, and like circumcision may have been going on forever. Ötzi, who died and was mummified in the European Alps five thousand years ago, bore fifty-seven tattoos, mainly over joints in which he had developed arthritis. Ötzi's work was done by rubbing soot into small cuts, but various cultures have devised multitudinous methods. Some Inuit used a sooty thread, which they stitched through skin with a needle. Tattoo artists in the Pacific islands often employed a comb made of bone needles, which they dipped in soot and oil, then tapped through the skin.

Tattooing in an era without antibiotics was sometimes a fatal undertaking. But then, so is breast augmentation and liposuction, even with today's sterile tools and robust drugs. Whatever urge drives my species to alter the body, it's so compelling that humans undertake it with the full knowledge that they may not survive.

We've reached the far end of our animal. And what a peculiar assemblage of features am I. Although humans are similar in size to other great apes, evolution has pushed my body in a longer, lither direction. I'm an upright critter, patchily furred, knobby with fat and bones, and topped with the high dome of a skull. This organization of flesh and bone is less muscular and more vulnerable to attack than that of our fellow great apes. What's more, my naked hide leaves me wholly unfit for life outside the tropics. Like the hippopotamus or the naked mole rat, I look like an experiment or a work in progress. I would rather look like a leopard, who moves like silk and has fur like spotted velvet. Even a mouse, with a teardrop body and wide, nocturnal eyes, is a more pleasing package.

One benefit to this uncommon anatomy is a body that can carry me for miles at a time. So if I'm a little light in the weaponry, I can at least outrun some of my predators and chase down some of my meals.

And should my legs and my flimsy claws fail me, I can always resort to the contents of that bony dome. Ungainly as it looks up there, the human brain can compensate for physical deficiencies that would defeat a leopard in a New York minute. No hunting weapons? I'll construct a spear—or an entire farm full of domesticated meat. No defensive weapons to fend off the bears? Let's build a fire—or a house. Too cold in the Arctic? How about stitching caribou hides into a second skin? And if the legs fail completely, we have this wheel we invented, and we'll use it to substitute for legs. And we'll build some sidewalks so the legless humans can still roll around to do their hunting and gathering. . . .

The combination of a light body and heavy brain undeniably works. My species is prospering. And if we look a little funny while we're doing it, we can change that, too.

2

CRAFTY AS A COYOTE:

THE BRAIN

In the human, the brain has evolved to dominate the animal. The organ consumes a large percentage of the body's energy budget. In exchange, the brain makes possible the human's tremendous tool kit, which at this time ranges from stone axes to melon ballers and a space station. In fact, as humans have developed tools that substitute for their own legs, arms, digestive organs, and even vocal cords, some injured individuals continue to thrive primarily as a brain. (Stephen Hawking, the great astrophysicist, is an example of this type.)

The brain is strongly influenced by sex hormones as it develops in the fetus. Mature male and female brains can perform nearly all the same tasks but may execute them in completely different ways. Even the fundamental architecture of gray and white matter differs between sexes. The intricacy of the brain's components opens the animal to myriad disorders, some of which also afflict one or the other of the sexes disproportionately.

This rather baffling organ is even inventing tools for the investigation of itself. Thus scholars of the human animal are now finding ways to access the hidden brain and characterize its behavior.

No normal fact sheet would dedicate a separate section to the animal's brain. If the brain of a mountain goat or a lemur were unusual, it would merit a mention within the physical description. This rarely occurs.

A GREAT, UNGAINLY BRAIN

Most of my organs are inscrutable. I've never laid eyes on my liver or pancreas. But the mystique of the brain is particularly piquing. From time to time I get so curious about the contents of my skull that I'm suckered into an Internet IQ test. On the Internet I'm quite a genius. But I'm not smart enough to remember that the primary goal of these websites is to generate traffic, which is not going to happen if they tell visitors we're merely average. I suspect IQ inflation is rampant on the Internet. So I'm left with the firsthand observations I can make of my own head: It excels at detecting patterns. It stinks at holding more than a few elements: (3 x 9) x (2 x 17) + (79–97) equals . . . what?

I'm most often aware of my brain when I encounter its limitations or malfunctions. It grinds to a standstill after six or eight hours of hard thinking. When a migraine tinkers with its blood supply it wanders slowly, picking up an idea but failing to take it anywhere useful. And deprived of serotonin-boosting drugs, it will seize on any violation of justice, chewing the sore spot obsessively. So that fur-covered dome behind my eyes houses a temperamental and self-analyzing, well, person.

It takes up a lot of space, this person/brain. And it's hot. This pillow of fats and proteins is so fraught with thought that it's smoldering. My brain alone burns 20 percent of my chow. The rustic brain of some marsupials, by contrast, burns less than 1 percent of the calories those simple animals eat. My wad of wiring is so hot and bothered that it puts all the world's other brains to shame. Or does it? Much has been made of my colossal cranium, but it is not nature's biggest. Whales, elephants, and dolphins have bigger brains. Mine isn't even the biggest per body size: Hummingbirds and a Central American squirrel monkey apparently trump me there. Nonetheless, my brain is admittedly, undeniably, monstrous compared to most other animals'.

Clearly, size isn't everything. Consider that my own female brain is smaller than my mate's, yet it seems to work just as well. In fact, size is only one variable in how my brain differs from my mate's—and my neighbor's, and yours. As intricate as the human brain is, it's surprisingly easygoing about the organization of its innards. But size does demand attention. The human animal demands we pay attention to the differences between one of us and the next. And our species has been chewing over the differences in brain size and intelligence for a long, long time.

The history of the huge human brain is like that of a balloon inflated by a series of breaths spaced over a few million years. Our early ancestors, like little Lucy (*Australopithecus afarensis*), had brains the size of modern chimpanzee brains—a large orange. That was 3 to 6 million years ago. But by about 1.5 million years ago, *Homo erectus* carried a brain nearly as big as mine. It carried him, too, and produced a tool kit of surprising sophistication. The modern scientists who re-created the island-hopping lifestyle of *H. erectus* in Indonesia had to invent Stone Age Tupperware for food and water, and build a sailing raft worthy of rough seas. The brain of *H. erectus* must have invented all those tools, too. It was probably *H. erectus* that next begat that stocky mutation, *Homo neanderthalensis.* Cousin Neanderthal, who roamed from Europe to western Asia until thirty-five thousand years ago, still holds the record for hominid brain size. Individual brains vary—even today, adult brains measure from one thousand cubic centimeters in Pygmy populations to twice that in the fattest heads of our species. But on average, *H. neanderthalensis* topped *H. sapiens* by a few percent. *H. neanderthalensis* had a bigger body, too, so this big brain wasn't out of scale. But in any case, Cousin Neanderthal and his big brain went extinct. Our line endured.

(Both species coexisted for a while in Europe. The tantalizing possibility exists that my forefathers and -mothers were similar enough to the Neanderthals that, like horses and donkeys, they could interbreed. This would have twisted a strand of Neanderthal genes into *Homo sapiens* DNA, especially in Europe. I love the idea that I could be part Neanderthal. It makes me feel exotic.)

The survival of my species shows again that size isn't everything. Other animals illustrate the same point. The New Caledonian crow has a brain the size of a small pecan, yet the bird can spontaneously clip and bend leaves, sticks, feathers, and even wire and cardboard, to make probing tools. British scientists who work with Betty, the most famous of these birds, have concluded that her intellect is on a par with that of the chimpanzee, whose brain is dozens of times larger. I, having access to a local pack of dogs that range from a one-hundred-pound Bermese to a six-pound puggle, could have told them the same thing: A bigger brain is no guarantee of greater sapience.

But is this true among humans? For nearly two centuries scientists have haggled over the relationship between human brain size and smarts, but have forged no clear connection between the two. Still, the subject inspires both scholars and outrage. There simply is no socially acceptable way to ask the following questions: Are the smallest humans less brilliant than the biggest? And are those with average skulls condemned to be ordinary? Only scholars of humans get anxious over such questions. Scientists who study crows and chimps don't experience such angst. It's not at all rude to investigate whether the biggest-brained bird is the smartest. But among *Homo sapiens*, burdened with a tradition of dividing and conquering one another, it's a wince-inducing issue.

It was less so back when European races began to question the morality of their treatment of others. It helps to explain why Friedrich Tiedemann was able to collect a bunch of empty skulls and fill them with millet seeds. This German anatomist's seedy measurement of brain volume made an impression. Tiedemann was weary of excuses for slavery, and in 1836 published science—real measurement and comparison—demonstrating that the skulls of Europeans, Ethiopians, and a whole bunch of other races accommodated the same range of brain sizes. Finding no racial differences in brain volume, he would admit no inherent difference in intellect.

But Tiedemann's millet didn't put the issue to bed. He merely demonstrated that all races share a range of brain sizes. Humans still wanted to know whether those brains at the smallest end of the scale, of any race, thunk the smallest thinks. Since then science has advanced from weighing millet to measuring hat size to MRI pictures, to no avail. Even the most stringent and technological attempts to pin IQ to brain size cannot overcome the complexities of the subject. In 2005, headlines ricocheted around the world when an American psychologist published a meta-analysis of some twenty previous studies. The data, mixed and matched and massaged, satisfied him: Smaller brains produce smaller IQs. But still the issue won't lay down its head.

For one thing, the difference he reported was teeny-tiny, far too wee to show up in a small sample of humans. What's more, measuring brain volume is hard. Even the high-tech methods like MRI are fraught with problems. And if you think measuring a brain is tough, try measuring intelligence. Try measuring it in a nonliterate culture, like the Yanomami of the Amazon. And then, what are we to make of studies that link IQ to the size of specific lobes and pieces of the brain, as opposed to the whole cabbage? Finally, it's possible that many studies of brain size and IQ have accidentally measured the effects of a deleterious childhood. Hunger and disease can stunt a body, and independent of that, they can also stunt an intellect. So it's possible that a small-headed human with a healthy childhood has no IQ deficit, but when you add a disease-stunted body to the mix, his stunted intellect drags down the average. I, in the company of many, remain unconvinced that a small brain is an intellectual liability for the human animal.

Still I am obliged to ask this next question, which is even ruder than the last: How do *really* small brains, like the brains of Pygmies, compare to mine? I'll delay with some housekeeping: "Pygmy" is a generic name, and not always a welcome one, for races whose height averages less than fifty-nine inches (that's four foot eleven). Various Pygmy populations are scattered throughout Africa, Southeast Asia, New Guinea, and the Philippines. I share their body proportions; I'm just much bigger. And so is my brain.

The fretting over this question went public in early 2005 with an announcement about *Homo floresiensis*, the extinct Indonesian "Hobbit people." I opened my fresh issue of the journal *Science* to discover that the skull of *H. floresiensis* displayed signs that it was wired on a par with those of its progenitors and/or neighbors, *Homo erectus*. The brain volume of hulking *H. Erectus* was 980 cubic centimeters; that of *H. floresiensis* was 380 cc. But when Florida brain scholar Dean Falk used CT scans to model the brain of *H. floresiensis*, she detected outsized frontal lobes (planning and problem solving); temporal lobes (memory, most likely); and lunate sulcus (sensory analysis). This brain had shrunken, Falk proposed, but had maintained the fine circuitry that only modern humans possess. Falk suspected that *H. floresiensis* was equal to any toolmaking task that big-brained *H. erectus* could handle.

Dean Falk seemed like the kind of human I might ask about the brains of Pygmy humans. She was. "There is no reason to believe that Pygmies are less intelligent than other humans," read her blunt reply. (To generate hard data on this, you'd need an IQ test appropriate to the forest lifestyle of Pygmies.) She also corrected the record on body proportions: Pygmies, who presumably evolved from larger races of humans, do not have brains proportional to mine; they're actually bigger, relative to body size. Not that I expect it would make much difference if they weren't. As long as the correct lobes and fissures are present, size just doesn't seem to be very important when it comes to what a brain can dream up.

Why, then, did the human brain get so big in the first place? Why does my skull dome up over my brow? Why does my noggin take such a bite out of my energy budget?

The how is simpler. Evolution is built upon the copying errors that occur constantly in DNA. Any mutation that gives an animal an edge over its competitors is more likely to be passed down through the generations. That's how my brain got so big: Those ancestors who mutated their brains upward a size fared better than the competition.

But why did they do better? Did each puff of size add voltage to all the animal's abilities? Or was each inflation dedicated to a single ability, like communication, or foraging, or throwing weapons? Because the question is compelling, and because humans are creative, there's a theory to suit every taste. Remember, this bigness is relative—whales and elephants have bigger brains than mine. What I have is a brain bigger than you'd expect for my size. Usually, mammal brains are proportional to their bodies: A big body needs a big brain to manage its muscles, organs, and nerves. But my brain floats high above that scale. The company up here includes the other great apes, hummingbirds, seals, and the diminutive elephant-nosed fish, whose brain makes up 3 percent of its body (mine is 2 percent). So, whence my swell cerebellum? Here are a few theories:

Machiavellian theory: I want my neighbor to walk my dog with his today, so when I talk to him I'm going to mention my busy schedule. At the supermarket, I've got to avoid that clerk who demands ID when I buy wine, because I've misplaced my license. Oh, here's an e-mail from a powerful colleague who always requests help but never offers any—what's it worth to stay on her good side? It's a fact that the species with sneaky lifestyles have evolved the biggest brains. If your legacy relies on tricking your peers out of food or stealing a peer's partner for a quick mating, you'll need a lot of brain cells to keep track of who's who, who's where, and whom you can't trust. A recent experiment with various primate species concluded that the cheatin'est species had the biggest brains, per body size. The sweeter spin on this theory is the social intelligence hypothesis, which hangs its hat on the cooperative relationships that social species maintain, instead of the competitive ones. Case in point: I must remember to buy Amy lunch because she paid last time and will stop being nice to me if I don't reciprocate. Either way, it tends to be the case that species who live in groups and have complex social lives have big brains.

Foraging hypothesis: Having traveled widely, I've managed to procure food even in cultures where I speak not a syllable of the language. At markets and restaurants I investigate with my senses, or communicate with pantomime, or even sketch the food I desire. And again, a survey of other animals finds that those creatures who inhabit a challenging environment have evolved bigger brains that permit innovative survival behaviors. Birds who exploit the food-poor Arctic not only sport big brains but also demonstrate the highest rate of innovative behaviors. And a study of birds that humans have introduced to new environments found that big-brained species like parrots are much more likely to adapt and survive than pinheaded creatures like pheasants. So in birds, at least, a big brain helps to master an unpredictable habitat. Given the human animal's propensity to wander, which we'll explore in chapter 4, such a feature would have been valuable in our prehistory—the kind of mutation that gets traction.

Cognitive mapping theory: Clark's nutcracker, a midsize bird from the American Rockies, will store a winter's supply of seeds in some six thousand locations, then retrieve each one as needed. How? By holding a mental map in her almond-sized brain. Because my own cognitive maps incorporate a house full of stuff, a town full of more stuff, a planet loaded with important stuff, and even a universe that, while sparsely furnished, is nonetheless large, my maps are far more detailed than a nutcracker's maps, and thus take up a lot of room. Perhaps those among my ancient ancestors who kept better mental maps thrived at the expense of their comrades. Some folks think the superior spatial skills of the human male—and they are undeniably a bit better than a female's, on average—may account for the male's larger brain.

And finally (well, not finally, because these theories are abundant), my own, personal, you-heard-it-here-first theory: In the mirror before me one of the few features I truly admire is my right shoulder. In my youth, this sleek hemisphere of tendons and muscle, wrapped around a joint of bone, carried me to the championships of throwing—shots, javelins, discuses. But it takes more than muscle and bone to put a shot with any accuracy. So try this: The ancestor of modern chimps and humans occasionally used a stone to smash open food. When the hominids split off from the other apes, they began to use stones to smash the occasional enemy. This paid handsomely, as it stopped the enemy before he could sink his teeth into the stone thrower's face. Those whose enlarged brains could coordinate the act of throwing with precision lived long and reproduced. Through eons, the stone wielders evolved to balance on two legs, which allowed them to hurl stones with much greater force. They evolved further to smash prey animals with rocks. Mmm, brain food! The brain inflated, the rocks grew sharper, the aim grew deadly, and the canine teeth shrank to the obsolete souvenirs smiling back at me today. (Over on the other side of the ape split, modern chimpanzees will rise on two legs to throw with tremendous power, but their aim is deplorable. Long canine teeth are still their weapon of choice.) In keeping with theory fashion, I'll call this the theory of the Throwing Ape: A brain and body that could throw accurately begot an upright animal who could hunt a rich supply of brain food.

And . . . you may add your own theory here. The huge human brain today serves so many purposes that you could pin its evolution to almost any of its functions.

A SEX-SPECIFIC BRAIN

If my own brain is enigmatic, my mate's is more so. That organ I can analyze only via the behavior it produces, along with the strange insights science is yielding. Not that the behavior itself isn't strange. My mate's most obvious foible is that his brain can juggle a large number of elements simultaneously—but not as many as he thinks it can. Habitually he loads on the tasks—he's cleaning the woodstove, he's pondering a professional problem—and then reaches for one more. "How was your day, son?" The son's brain is familiar with the father's brain, and the son accuses the father of not listening. Half the time, the father repeats the boy's sentences verbatim, and takes umbrage. The other half, the father prepares to take umbrage, opens his mouth, and can't recite a word.

There's nothing specifically "male" about that behavior, of course. Very little of the human behavioral repertoire can be traced to sex differences in the brain. And that's weird. How can two brains be so different in their layout yet produce such similar behavior?

The brain in my skull is quite a different organ than the one in my mate's head. In fact, as I investigate this, I'm increasingly convinced that the human animal has two brains, each built, wired, and operated according to its sex.

For starters, the male brain is generally larger than the female brain. Throughout my species, females are about 15 percent smaller than males, with correspondingly smaller brains. This presents us with a mountain of data whose message is unambiguous: Females have smaller brains than males but IQ tests say those brains produce equal wit and wisdom. And if anything, it's the male brain that shows a vulnerability to size. To harken back to the size debate for a moment: When studies do tease out a link between size and intelligence, they sometimes also show that a small brain impacts a male's IQ more than a female's. The female brain may be less sensitive to smallness. Strange, eh? Again, we're talking about minuscule IQ differences, but any detectable difference hints at divergent design principles at work in male and female brains.

Some of those differences are written pretty boldly in the brain's physical structure. Richard Haier, an American psychologist, is one of the scientists now investigating human brains as though they sprout from two different animals: females and males. Using MRI scans, he measures the deviations in gray matter and white matter between the two sexes. Gray matter is the thought-generating layer coating the brain; white matter shuttles thoughts between different surface areas of the brain. And in those regions of the brain that support "general intelligence," Haier found, females have nine times more white matter than males. That's nine times more of the stuff that ferries data around the brain. As for gray matter related to general intelligence, males have six times more of the data-gathering stuff. These are huge disparities. They make a huge difference between my brain and my mate's. How can we be wired so differently—yet both build bookshelves and write thank-you notes?

Haier has also compared IQ to the size of discrete subsections of the brain, versus measuring the entire blob. And Haier's own brain concluded that the bulk of the human animal's IQ may be based in a handful of areas that govern memory, attention, and language. You could have a head the size of an ottoman, but if those critical sections are small, you're going to be a bit dim. Conversely, if you have a brain like a doorknob but you devote most of the knob to those three tasks, you could be a brain surgeon. Furthermore, Haier found disparities in where males and females store these crucial brain segments: I, as a female, hold my "smart matters," both gray and white, largely in my frontal lobes, where emotions, speech, reasoning, judgment, and movement dwell. Males, by contrast, strew their wits all over the brain. My mate has crucial patches of gray matter in his frontal lobes and nearby parietal lobes, where reading and math reside. His "smart" white matter resides in a completely different nation, the temporal lobes, home of sound perception and memory processing. With brains this different, I'm impressed that we can communicate at all, let alone agree what color to paint the den.

Now, Betty the crow has already instructed us that there's more than one route to cleverness. Her brain is small and smooth as a grape, while a chimp's is deeply crinkled and baseball-size—but the two brains cogitate comparably well. The surprising thing about the human brain is that two unlike brains occur within one species.

Granted, some degree of divergence should be expected. Other animals show a girl-brain/boy-brain dichotomy. For instance, many mammal brains contain a "sexually dimorphic nucleus," a buried clump of cells that's much bigger in the male brain than in the female brain. (Its function is unknown.) Many also host a sex-specific clump that dictates breeding behavior, as do my brain and my mate's. What's so puzzling about the human head is that the sex differences spill into regions of the brain that have nothing to do with sex-specific tasks—the parts governing speech, reasoning, and movement. The human brains are different in the sophisticated areas, where highfalutin thoughts are constructed. Why should that be? Why isn't the same thinking gear good enough for both sexes?

The answer probably lies way upstream, when sperm and egg first met. A 2007 survey of mouse brains found that 14 percent of the genes regulating the brain behave differently in male and female mice. If this applies to the human animal as well (and such things often do), the brain genes in my mate and me may be nearly identical, but they express themselves—build brain tissue—differently. And the lever pushing those genes around is probably our respective sex hormones.

When I was conceived, the DNA of my parents combined in a uniquely me mixture. I, master of my new-minted genome, decided how much testosterone and how much estrogen I would generate, and then I manufactured them. In that bath of sex hormones, my brain absorbed a uniquely me flavor.

All human brains form thus. The fetus forges a unique ratio of testosterone to estrogen, ranging from very male, which is rich in testosterone, to very female, rich in estrogen, and all shades in between. The recipe we each adopt flavors our brain, and thus our personality. You might wonder how scientists can tell if I brewed up oodles of testosterone or estrogen in the privacy of the womb. Scientists, too, spent many years wondering how to gauge this without poking mothers in the belly with a big needle. And then a UK scientist winkled out a clue, mounted on the outside of adult humans, which can testify to that time in the womb. John Manning discovered a hormone meter attached to the human right hand: The right ring finger, due to a quirk of genetics, grows longer under the influence of testosterone. My mate, who perfumed his fetal bath with testosterone, boasts a ring finger longer than his index finger. I, who scented mine with estrogen, have a ring finger the same length as my index. Had either of us dosed ourselves harder with testosterone, our ring fingers would tell on us. They would be longer. Had either of us produced less testosterone, our index fingers would have overshadowed the ring fingers. And as the ring finger grows, so does the brain: So here on my right hand is a dipstick displaying the strength of the hormonal marinade that gave my brain its sexual slant. Researchers can now use that finger to tease out associations between fetal hormones and adult brain styles. (Scientists actually observed the difference in male and female finger ratios over a century ago, but haven't known what to make of it.)

So what has testosterone done to my mate's brain? Freakish, fabulous stuff. Running my fingers through the scientific literature, I compile a list of traits you may be able to predict from a male's high-testosterone ring finger. But be calm: This stuff is statistical. Not every male with a supersize ring finger was soaked in extra testosterone. Some additional factor also must influence human finger length. However, statistics show that hormones do have an impact. (Another caveat: Before you check your fingers and freak out, note that real scientists determine length by photocopying the right palm, then measuring from the fingertip to the crease nearest the palm. The result is quite different from what you'll gather with a panicky glance at the back of your hand.) So, the associations, to date, with a male's elongated ring finger:

- He's more likely to be aggressive.
- He's more likely to mate with many females.
- His sperm count may be high, and the individual sperm cells may be healthier than most.

- He could be adroit with his left hand, even if he's right-handed.
- He could have a higher level of testosterone circulating in his body.
- He could excel at soccer. (The scientist who conducted this study analyzed the fingers of European soccer players. The result should apply equally to cricket, American football, and Ping-Pong.)
- He may face an increased risk of autism or Asperger's syndrome, of depression, migraine, stuttering, and schizophrenia.
- He may excel at math and stink at reading. (Contrary to the infamous proposal of former Harvard president Lawrence Summers, testosterone appears to have no impact on science aptitude. Nor does mathematical aptitude predict success in science.)
- And he's more likely than average to be gay.

Extra testosterone does indeed correlate with homosexuality. Which presents a good excuse to repeat the caveat: These are statistical correlations. A male with a long ring finger isn't automatically gay, or a lady killer, let alone both. It's just that if you take one thousand males with long ring fingers and one thousand with normal ones, then among the long fingers you'll find a higher percentage who are gay, *or* who are aggressive, *or* who are soccer phenoms.

On the other end of the spectrum, males with unusually short ring fingers—hence low-testosterone brains—have their own peculiarities.

- They may have elevated lactation hormones.
- Their sperm cells may be dead on arrival, if they exist at all. Egad! (This is just statistical, remember? Just an elevated chance, not a biological decree.)

And now for the females:

- They, too, tend toward aggression and homosexuality if they show long ring fingers.
- They also enjoy that left-handed dexterity if they're long in the ring finger, along with some of the mental-health hazards.
- On the other hand (heh!), if they have *short* ringers (indicating very low testosterone), they confront a higher risk of breast cancer.
- Short ring fingers also predict that they're more likely to be left-handed, and quick to retrieve words from memory.
- They could have higher levels of the hormones that regulate ovulation and lactation.

As fascinating as these lists are, their relevance here is that these physical features allow us to peek into the human brain. The point is that this hormone-poached dumpling atop my neck is as much a sex organ as are my ovaries and mammary glands. My brain and my mate's look similar from the outside, but the filling is sex-specific.

A LOPSIDED AND SCRAMBLED BRAIN

Before abandoning the brain case, I need to exploit one more physical feature that affords a glimpse into the brain. Looking back at my hands, it's the right one that leaps from the keyboard to the mouse to the phone to the pen, doing fifty times more in a day than the left one does. The left half of my brain is "dominant," and it transmits that bossiness to my right hand. But in roughly 10 percent of developing fetuses, the right half of the brain assumes dominance. As these humans mature, their brain dominance becomes evident in their aggressive left hand. So left-handedness—just like a long ring finger—gives scientists a gauge for measuring a brain's hidden qualities. If you take one thousand lefties and compare their mental strengths, weaknesses, and anomalies to those of one thousand righties, you can tease out more of the brain's secrets. The correlations are many, miscellaneous, and at times confusing:

- Males account for 55 percent of lefties; females 45 percent.
- Female lefties face a doubled risk of developing breast cancer before menopause.
- Homosexual males are somewhat more likely than heteros to be left-handed.
- Homosexual females are much more apt to be left-handed than hetero females.
- Humans with autism (80 percent of whom are male) are more often left-handed than their peers; schizophrenics may also lean left.
- Lefties are more inclined to immunity problems, including childhood asthma.
- They have a higher risk of dyslexia and stuttering.
- Male lefties, but not females, are more likely to be mathematically gifted.
- Children born deaf are twice as likely as their hearing peers to be left-handed.
- Lefties have fewer offspring.

Two messages waft from this data pile: Right-side brain dominance impacts males and females differently. It also seems to scramble some of the brain regions, affecting language and math performance.

A third observation might be this: With all those challenges to their health and reproduction, how have lefties managed to persist in the human population over time? And they have persisted, judging from one scholar's analysis of cave paintings. In the days before canned spray paint and graffiti, humans expressed their artistic impulses by pressing a palm to a cave wall and spitting paint at it through a hollow grass stem. This hand motif was very popular in the European upper Paleolithic (ten thousand to thirty thousand years ago), so French evolutionary ecologist Charlotte Faurie and her adviser Michel Raymond were able to find 507 prints to analyze for handedness. If a print was of the left hand, they reasoned, then the painter had used the right for the delicate task of aiming the straw. Then, they convinced seventy-nine of their late-Holocene students to make modern hand stencils. The modern *H. sapiens* produced a ratio of lefts and rights identical to that of Paleolithic cave painters. Conclusion: The percentage of switch-spitters has been stable for at least thirty thousand years. So left-handedness must confer some advantage.

That advantage may be a superiority in fisticuffs. If I, a solid righty, am accustomed to fighting humans who are also righties, then when I face a lefty, I'm going to get pegged. The blows are going to come from a direction I don't anticipate. The lefty is going to punch my lights out, steal my mate, and breed happily ever after, passing along her successful genes. A glimmer of data supports this hypothesis: Lefties heavily populate baseball, cricket, tennis, boxing, and fencing. Their unexpected attack from the left is, in a sense, deadly.

But of course, there are other theories.

I also like the theory that addresses the unorthodox floor plan that often appears in a lefty's brain. In most humans, the brain's various functions—speech, vision, memory—occupy predictable locations in the left hemisphere or the right. But in lefties, these brain functions often seem to have been tossed in the air to fall where they may. This fresh arrangement (the theory goes) can produce wild artistry, fresh philosophy, and flashes of genius, in addition to the hazards listed above. And indeed, musical talent, at least, does seem to crop up more among lefties than righties. Of course, lefties may persist for both reasons, because they're successful in both combat and life.

Handedness is not unique to the human animal. My human brain is strongly lateralized, with speech production on the left, shape recognition on the right, and so forth. But many other animals tilt, too. Most rats make right turns when they're in search mode. Many species of tadpoles are "left-eyed" when they approach another tadpole (having a sphere of head between their two eyes, they have to choose one or the other). Walruses appear to be right-flippered in the same proportion that humans are. Each New Caledonian crow that uses sticks to "fish" for worms shows a clear handedness, but just as many are righties as lefties. Ditto for the manipulations of gorillas. The handedness of chimpanzees is a special case. Some scientists see a bias toward the right; others argue that right-handedness only appears in captive apes, not in the wild. An illuminating result of the argument is this observation: Of those chimps who fish for termites, the ones who specialize with one hand catch the most.

And therein may lie the genesis of human handedness. It may be that focusing the brain's efforts on one hand, rather than dividing it between two, yields greater efficiency—in feeding, child rearing, and fighting. It may also pay to avoid a split-second's indecision over which limb to deploy in an emergency. After all, when an enemy is running at you and a stone lies before you, you don't want to squander time deciding which hand will hurl the stone. Science can't yet say why 10 percent of humans come out left-handed, and they know even less about why we're handed at all. We, like many other animals, just are.

I sometimes wish I could reach in and touch my brain—shake hands with it, if you will. Although it operates my entire body and personality, the organ is hard to know. It's a little ironic how much human brainpower is currently bent on making sense of the human brain. The phenomenon puts me in mind of trying to view the back of my own head in a mirror. But how could we ignore such a feature? The brain, with its unequaled complexity, not to mention its impulse to examine itself, must be the component that sets humans farthest apart from all other species.

And it may be evolving even today. All it takes, after all, is a helpful typo in the DNA—a copying error that alters the meaning of a gene. And according to Bruce Lahn, a Chicago geneticist, the brain-size genes in primates have been flubbing at a frightful clip for at least 30 million years. He deduced this by comparing humans to rats and mice, whose brain genes have been mutating at a more relaxed pace.

He then identified a couple of genes inside which mutations occurred at rather noteworthy points in human history. One of the mutations appeared on the *Microcephalin* gene about 37,000 years ago, which is the midmorning if not the dawn of "culturally modern humans"—humans who made musical instruments and art, and buried their dead with ceremony. This was evidently such a beneficial accident that today it's reproduced in 70 percent of all human brains, most commonly in Europe and eastern Asia, but is also scattered throughout the human population because of our propensity for travel. Then a second gene, *ASPM*, sustained a beneficial mutation just 5,800 years ago, as agriculture and written language were revving up. Exactly what these genes do isn't clear. The reason Lahn thinks they involve brain size is that when a glitch turns them off in a developing fetus, the brain fails to mature. Lahn also thinks these two accident-prone genes are just a start. He guesstimates it has taken hundreds or even thousands of similarly successful mutations to create the modern human brain, and that a million years from now the organ will be considerably different from the one hidden under my head fur today.

Somewhere on this planet, at this moment, a fetus may be growing with a mutation in her head that increases the speed at which her brain works. I leave it to her to draw any definite conclusions about the human brain—its size, its sexuality, its split personality. My best effort at a conclusion might be this: The brain, mine and other animals', excels at flexibility. Some creatures, like the New Caledonian crow and *Homo floresiensis*, can get a huge amount of work out of a small brain. My own female brain is arranged so that it functions on a par with a male's bigger one. And even the world's larger brains, like those in elephants and whales, don't appear to produce the bonus of deep self-awareness and innovation that I, a human, display. It's a befuddling organ. In terms of getting its host through the day, the brain of the simple kangaroo or possum does the job with a lot less confusion.

3

BLIND AS A BAT:

PERCEPTION

This animal perceives the world foremost with its eyes. As in many predators, the eyes are forward oriented. This produces three-dimensional vision but narrows the total field. Inside the eye, the human (like many of its fellow primates) has three cones rather than two, producing color vision much richer than that perceived by most mammals. This animal's visual system also incorporates a sensitivity to the "body language" of conspecifics—and presumably of other species as well. In fact, merely viewing a pair of eyes widened in fear serves to stimulate a human's own fear. Vision is by far the most acute sense.

Taste and smell, the chemical senses, are rather weak. The tongue can, however, identify poisons with passable accuracy and is sensitive to high-calorie sugars and fats. Recently the nose has been found to draw crucial information from the sweat of other humans. Hearing is acute, as it often is in animals with a narrow field of vision; the ears alone can "point" the human head toward a sound with an accuracy as fine as that measured in any other species. The ear also can capture a great range of frequencies, from deep basses to squeaky trebles, a noteworthy feature given the restricted bandwidth perceived by many other creatures.

The sense of touch has received little scientific attention, but regular stimulation of this sense appears important to the normal development of young (as has been shown in laboratory animals, too). Females demonstrate a slight edge over males in the speed at which their brains process sensory information—especially during the fertile phases of their menstrual cycles. The human may be capable of perceiving its environment in other ways, as well. One might ascribe to them, for instance, a sense of balance, direction, or time. And many humans believe, although data is lacking entirely, in a "sixth sense" that would explain such common experiences as prescient dreams, mind reading, future telling, and assorted hocus-pocus.

EYES

For most of my childhood our family included a little screech owl. Wowl had arrived as an orphan and matured to ride around on a human shoulder or watch over the household from atop the grandfather clock. From time to time we'd notice him staring across a room, fixated on something we couldn't see. We kids would make a game of it, sort of "warmer/colder with owl." Guided by his gaze, we'd try to locate his target. Eventually, with our noses nearly pressed to the plaster wall, we'd happen on a fruit fly at rest or a tiny spider.

When I look out my office window to the backyard, it seems that I'm seeing my environment pretty clearly. The finest branches of the tree by the deck are crisp. The green garden hose leaps out from a background of brown leaves on the lawn. Farther out, I can discern the gray peaks of a picket fence, even against a dead-drab lawn. And beyond that, the neighbor's treetops . . . aren't really clear at all. Wanting in detail, admittedly. Truthfully, kind of a fuzz. If Wowl were here on my shoulder, he could probably detect the yawn of a chickadee perched on one of those fuzzy branches. Evolution has awarded to each animal that light-detection gear most suited to its particular needs. I don't need to see chickadees yawning in order to make it through another day. A *Homo sapiens* has her own needs.

The eyes of all animals had a humble beginning. Some single-celled somebody probably got the vision thing rolling when a DNA flub granted it a light-sensitive chemical. This dab of chemistry may have allowed an ancestral Mr. Microbe to eat better, or to better avoid being eaten, and thus he thrived. And everyone who evolved from him was grateful for his revolutionary photopigment.

Today, each species packages its photopigment a little differently (or a lot differently), but the general rule is the same: You collect a subset of the light spectrum and use the data to plot your next move. You don't even require a brain to do this—the marine brittle star is paved with crystals that detect a predator's shadow and signal its five arms to scramble for shelter.

I'm always surprised, when I pull back my eyelid, to see how big the eyeball is. The part that normally shows—the blue iris and a wedge of white on either side—is modest. (A little too modest for my sense of beauty, but I suppose that's asking a lot of a sensory organ.) As I wander the world, my top and bottom lids are open less than a half inch. But tucked into my skull is a slippery white sphere an inch wide. Eww. I admit I'm a little unnerved by my eyeballs. I can recall standing before the mirror as a teenager, staring into my own eye until my intellect ran aground on a shoal of identity issues: So, this is my eyeball looking into another eyeball, which is myself—or itself? It seems to have a cool presence of its own. Am I looking at myself or into myself? Brittle stars don't suffer such anxieties. They measure light and act accordingly.

Eye Style: Spherical

My human eye is unusual in that it really is a ball. I lift the blanket off the head of my dog, Kuchen, who sleeps at my feet, and study the eye he rolls up at me in annoyance. His looks bigger than mine, though his lanky body weighs in at half my size. But when I research canine eyes I find they're smooshed, front to back, like a beanbag chair. The eyes of the sparrows in the yard are shaped like pears, stem end to the front and fat end inside the skull. A scallop contemplates the sea through a number of half-sphere eyes that hang along the edge of its shell like a string of empty soup bowls. In a shark, the bowl is inverted: The bulging side gathers light rays and forwards them to a flat retina. The fruit fly's conical eye is set in its head like a diamond in a bezel. Many animals, including lizards, even have a flattened third eye, complete with lens and retina, under a special scale on the top of the head. It can't form images, but it does monitor light levels. Perhaps most unnerving to an eyephobe like me is the "pit eye" found on worms and mollusks. As you might guess, it's an open pit sunk in the animal's skin, lined with photoreceptor cells. Again, eww.

Scanning this world of eye styles, it occurs to me that most of these animals will never know the luxury of a roving eye. Pit eyes are obviously not able to cast themselves about. Nor are the compound eyes of insects, which are part and parcel of the head. The sparrow's pear-shaped eyeball will never roll as mine can, because a collar of bone secures the pear's neck. I now grasp that some of the most characteristic movements of animals—the cocking of a bird's head, the tipping of a dragonfly's—result from a need to move the whole head in order to move the eyes. Owls, I discover, are among the most eye-bound. Their eyes are greatly elongated, flaring like bells inside the skull. Locked in bone sockets, they take up more skull space than the owl's brain. The huge eyes generate crisp images, but they're immobile. Wowl, to shift his gaze, shifted his head. I remember his mannerisms so clearly: He'd rotate his head, then nod up and down to adjust for elevation, and finally the yellow irises of his enormous eyes would widen or contract to accommodate the new view. I wonder if a rollable eye is one of those features that soften our species toward certain animals—that make them more "human." My dog's expressions do depend heavily on the way his eyes roll up at me in mock mournfulness, or rotate noseward to study a beetle.

But back to my own eye. It's a pretty straightforward affair, for a mammal eye. Light enters the cornea, which directs it through the pupil. Behind the pupil, a lens focuses it further (inverting the scene in the process). Each redirected photon smacks a light-sensitive cell in my retina. That cell sends a signal through the optic nerve to my brain, whose job it is to make sense of the data. My pupil can dilate and constrict, depending on whether it's in a sunny or dark environment. My lens can adjust to focus on a book under my nose or on the peak of Mount Washington seventy-five miles away. None of this is remarkable in the world of mammal eyes. My eyeballs themselves are very much the no-frills model.

The eyes of my childhood pony, Duchess, were another matter. As a kid, I gazed into those brown orbs and wondered, *Why are there mountains in Duchess's eye?* The horse's pupil was a long oval, parallel to the ground. And along the top edge of that pupil, rising toward me through the liquid of the cornea, was a row of dark buttes. Now I learn that the buttes were really awnings, and that lots of animals have them. Technically, they're "iridic granules," or the "corpora nigra." Functionally, they act as visors, shading the pupil from a glaring sun. Animals who evolved in wide-open spaces are most likely to have them.

I'm also missing a tapetum lucidum, a spooky feature with an enchanting name: It's Latin for "bright carpet." If you've seen the glow of an animal's eyes at night, then you've laid eyes on its tapetum lucidum. Nocturnal creatures and other low-light specialists have evolved a reflective layer at the very back of the retina. In dark conditions this layer bounces sparse photons back through the retina's photoreceptor cells for a second chance at making an impression. When your headlights blare into the wide-open pupil of a nocturnal raccoon, all those extra photons bounce off the tapetum and return to you. Different animals' eyes shine in different colors because they use different chemicals to weave their bright carpets. The red glow that appears in photographs of my own eyes is not the same. Unlike a raccoon's or a deer's, my pupil doesn't constrict in time to block out the flash. But the camera catches a portrait of my blood-suffused retina, not a bright carpet.

Furthermore, my iris is boring. It's a blue ring with a round aperture. Consider the blue-and-orange Tokay gecko: Her golden iris is scalloped on the inner edge, and when she damps it against bright light, it forms a ruffled, vertical slit. Or take the octopus, whose bronze iris is bisected by the dark smile of a drooping pupil. And the tropical five-lined cardinalfish emblazons itself with black and white stripes that run from head to tail, and continue right across the iris of its eye. Given the possibilities, the eyeball staring back from my mirror is pretty humdrum.

But it is a serviceable eye. My lens, for instance, is more powerful than my dog's. Whereas Kuchen can see a tennis ball in motion, he's hopeless once it falls into rough weeds. To my eye, it's now a sphere against a background of crisscrossing foliage. To Kuchen's, it's a bump embedded in a blur. He relies on his nose to track it down.

My retina is fairly fine grained, too, which produces a clear image. Photoreceptor cells pack my retina at a rate of 200,000 per square millimeter. The cane toad of the American tropics gets a grainier view of the world with just 46,000 receptors per square millimeter. The California ground squirrel gets a similarly coarse picture, and rats have truly ratty vision. Of course, some other animals make my 200,000 receptors look ratty, too. Those sparrows outside have about 400,000 per square millimeter, and many raptors cram in a million.

As for processing speed, my eyes are rather indolent. When I sit watching a movie, the series of still images blends together to form motion. But to a fly, a movie plays like a plodding slide show. As a fly watches a movie, an image triggers her photoreceptors . . . and it continues to trigger them in the same pattern . . . and finally, a new image appears. A fly's eye can process two hundred different images in a second. I can handle twenty before they start to blur. The images in a movie are timed to exploit this. My eyes don't register the individual still images, but instead knit them into a flowing scarf of action.

How fares my field of vision, then? A bit better than average for a predator, I suppose. Because my eyes point dead ahead my total field of vision is just 180 degrees. However, most of mine is binocular, giving me excellent depth perception over 120 of those degrees. This is a lot of stereo vision, even for a predator. Cats and dogs, both predatory animals with eyes facing front, have narrower fields of depth perception than I have (but greater fields of vision overall). My binocular vision permits me to measure the distance between me and my prey, even when I'm dashing toward the cookies. At the same time, my peripheral vision feeds me data about my surroundings. A naturalist's trick I learned in animal-tracking camp: Standing in a field I unfocused my eyes. The landscape, formerly motionless in the summer heat, leapt to life: At my feet a bumblebee busied. To my right a grasshopper ascended a stalk of goldenrod. Overhead two black crows crossed the sky. And to the left a butterfly frittered up from the weeds then plunged back in. When I was fixing my predatory eyes on particular points, I was unconscious of all this hubbub. But my eyes were monitoring half the world.

Mine is a predator's version of vision. Compared to a typical prey animal, I'm half-blind. The common snipe and the woodcock, two long-billed North American birds who probe in leaves and mud, both have a 360-degree field of vision. Their eyes are set so far back on their heads that the two view fields actually overlap in the *rear*, giving the birds a bonus slice of binocular vision behind the head. That's a little hard for me to imagine.

Prey animals, like the snipe, rabbits, deer, and other sitting ducks, sacrifice depth perception in order to mount their eyes on the side of the head, where they can keep a near-360-degree vigil. But some of them have evolved behaviors that compensate for the missing depth perception. Various birds, lizards, and rats pump their heads up and down, or back and forth, to reveal which objects in their field of view are near and which are distant. To test-drive "pigeon vision," I turn to the window, shut one eye, and bob my head. The vase on the windowsill "moves" more than does the wheelbarrow on the lawn, which in turn moves more than the distant fence. And as I bob and weave I suddenly remember the characteristic way my father used to tilt his head as he studied a landscape. I never put it together until now. Dad, following a freak mishap of dentistry, had gone blind in one eye.

As a rule, predators, with a narrow field of vision, don't compensate much. But we humans, always happy to wield a tool, have recently overrun the natural limits on what we can see. We have created lenses, mirrors, and cameras that can deliver us images from behind our heads or even from behind the moon. Two of these tools cling right on the flesh of my eyeballs, delicately warping light to compensate for those eyeballs, which are not, I must confess, perfect spheres.

Color Perception: Excellent

It's nice to see in 3-D, but it's lovely to see in color. And that I do exceptionally well. My office is a testament to that. Above the computer is a painting of New Zealand mountains rendered in shades of sapphire, orange, and emerald. Below it stands my clock, a shiny ceramic confabulation in yellow, orange, and blue, topped with a green and purple parrot. On the right wall hangs a pink and turquoise Chinese silk dog, dangling from a gold and scarlet tassel. And the mirror to my left is framed with zigzags of melon, green, and yellow. Color doesn't just stimulate my eyeball, it stimulates my whole self. My one-eyed father was likewise a color junkie. He used to write his microbiology lectures beneath a four-foot sun he had painted on his wall in trompe l'oeil shades of red and orange. His vice was so notorious in his town that the owner of the local junk shop set aside everything—dishes, wastebaskets, art glass—of a hue sufficiently, glowingly orange.

The back of my eye is paved with a layer of rods and cones, elongated cells that detect brightness and colors, respectively. The human eye is unusual in that it has three different cone classes, each tuned to a different wavelength, or color, of light. When photons bounce off the wheelbarrow on my lawn and strike my cones, the cones fire in a combination that my brain interprets as "red." When my cones snare photons from a clear sky, they send a different set of readings, and my brain says "blue." My brain can distinguish some seven million colors with this system.

In the mammalian world, that's a luxury. All but the primates have only two types of cones. One of them is usually tuned somewhere in the shorter wavelengths (green, blue, or purple), and the other to a longer wave (yellow, orange, or red). As a result, most mammals perceive a depleted range of colors and a washed-out world.

The same blunting affects a small and steady percentage of humans who are color-blind. In those retinas, one of the three cones is tuned to an inappropriate wavelength. Depending on the cone and the distortion in its tuning, the world appears watered-down, with many hues parading as the same shade of gray-pink or gray-green. (Simulations abound on the Internet.) This genetic trait is sex-linked, occurring in one out of every eighteen males (and just one in two hundred females). After decades of speculation as to why such an anomaly persists in the human race, new science has produced a clue: Color-blind humans can negotiate a universe of beige in which I am rudderless. When British experimenters showed humans color swatches containing two shades of khaki, humans like me gazed in bafflement. But the color-blind subjects proved to be savants of khaki, identifying fifteen distinct varieties. The researchers speculate that the color-blind gene has persisted because it gave hunters an advantage in spotting khaki-colored animals in the khaki-colored grasslands of human prehistory.

Birds, lizards, and fish are much better equipped than mammals for a colorful world. They often have receptors tuned to five different wavelengths, instead of my three. But how that would affect their perception of my office is hard to say. It's one of the mysteries of vision: Two humans may agree that the sky is "blue," but one human may be seeing something the other human would call yellow. There's no way to know.

Night Vision: Deplorable

When I send Kuchen into the yard for his final pee of the night, he rushes into the shrubbery roaring. I, wondering which urban mammal I'll find hanging off my dog's face, follow. I bang into the picnic table. I stretch out one foot, then the other. I hold a hand before my face, sure I'm going to get a stick in the eye. If there was any doubt, a quick count of my rods and cones confirms that I am no nocturnal animal. Some night creatures use almost no cones at all, and why they have even a few is a mystery. The benefit of cones is that they subdivide light into specific hues. Because this would dilute the faint light of nighttime, most nocturnal animals devote the retina entirely to rods.

I'm not actually hurting for rods—my retina holds many more of them than cones. And these rods perk up to a light one thousand times fainter than what's required to fire a cone. But all rods are not created equal, and human rods make fuzzy images. Their low status is affirmed by my fovea. This is a spot at the back of the retina absolutely jammed with receptor cells. It's the sweet spot, the part of my eye I aim at a splinter in my thumb or at a deer in the distance. It's the only part of my entire visual field that's really in focus. And it's almost entirely cones. When I look into the night sky to locate a dim star, any light that falls into the fovea vanishes. The cones, by dividing the light among themselves, dilute it to invisibility. I have to shift my focus sideways from the target, so that starlight falls on a part of the retina with more rods. No such limitations hamper the South American oilbird. A cave dweller who flaps forth at night to eat oil-palm seeds, this animal has 123 times more rods than cones. And those rods are exceptionally slim and tightly packed, so that the dimmest glimmer of photons can still strike a crisp image.

Still, I'm not hopeless at night—not as hopeless as a hawk. If I take time to adapt, my rods can pick out a tree or a rock even by the faint light of stars on a moonless night. Even on an overcast night, I'm not utterly blind, thanks to all those blurry rods—seventeen rods per cone, in fact. The red-tailed hawk has just one rod per three cones, poor thing. The crows who often visit the oak tree outside my window wield only two rods per cone. (One scholar has speculated that the reason crows flock to city parks at night could be that humans leave the night-light on for them.) The worst night vision I can find belongs to the Australian bobtail goanna. This lizard's cones outnumber its rods eighty to one. Unless those are spectacularly sensitive rods, that reptile is destined to stub its toes at night. If my rods outshine the lizard's, it's probably because my species has prehistorically hunted, or been hunted, at night. Lending weight to that theory, my ability to detect movement "out of the corner of my eye" at night is quite respectable.

Et Ceteras

And speaking of movement, on my research journey I stumble over an intriguing phenomenon called "biological motion." Apparently, my brain is particularly alert to any pattern of rod and cone firings that might add up to an approaching human. I kind of knew that, in retrospect. We've all had our attention drawn to the corner of our eye, where a fellow human was moving in an unexpected way. Many of us have visited dangerous settlements, where we screened each human for signs of aggressive intent. And from every passing human, we can gain information merely from the way the muscles and joints flow: The human is limping, is swaggering, is sneaking, is parading. And evidently, we register these biological motions automatically.

Scientists can test this by filming a human walking toward their camera, then reducing the image to moving dots: at the foot, knee, waist, shoulders, head, and arms. A human brain, watching video of the dots, has no question about what it's seeing. Research shows that the brain lights up in those areas that analyze another human's behavior and prime the body for action. Actually, the watchful human doesn't even require motion to gather crucial information from another body. The posture alone of a frightened human—face hidden—is enough to fire the viewer's emotion centers and prepare her body for confrontation. Simply human eyes, with face and body blanked out, can convey fear to another human. It seems that, like a starling who flies up from the ground the instant his neighbor flaps in alarm, the human has also evolved to perceive fear in her fellow humans even at a distance.

I would be surprised if further research didn't show that my brain is particularly sensitive to signs of aggression in an approaching human. Even modern humans, confined by the bonds of culture, can be dangerous; I imagine that in the pre-police ages, every stranger was cause for both interest and alarm. Our knee-jerk fear of strangers appears hardwired in the brain, tipping our suspicion toward humans who don't look like the others in our home community.

Now, as we observed in the previous chapter, the male and female human have quite different brains. And since it's the brain that makes sense of visual data, I'm led to question whether my mate and I differ in our perception of reality. And, of course, we do. First, there's the fact that color blindness favors males. More racy is the fact that the human female of childbearing age sharpens her vision when she's most fertile, in the days before ovulation. On those days, I will be more sensitive in dim light than my mate. But whether this is an adaptation that helps a female gather information for her mating choices, or whether it's a side effect of hormonal fluctuations, is currently anyone's guess. We'll see as we move through the other senses, however, that they all grow stronger as a female reaches that crucial mating moment each month. Given that each of the senses runs on different chemicals, the side-effect hypothesis might be a stretch.

Of all the senses, my eyes are most important to my natural history, my lifestyle on Earth. A sprawling tract of my cerebral cortex—at least one-third—lights up to process the visual data I gather when I cast my eye on my clock, the wheelbarrow, my mate. This leaves less room for the other senses. And all this for an eye that is a rather average instrument.

Although mine tunes in more colors than most mammals can see, it's less color sensitive than a bird's. It's weak at night. And if my visual brain can sift crucial information from another human's face and body, this is not a rare talent in other animals. With an adjustable lens behind my pupil, I can zero in on objects or zoom out to infinity—but the resolution is modest and the processing speed slow.

Studying my eyeballs again in the mirror, I consider the contact lens on my iris. I began doing "close work," as scientists call reading, at age four, and a few years of zeroing in can alter the shape of a human eye. By my early teens I was struggling to zoom out. In my culture, distance vision is the first major system to fail, often blurring even before a human reaches breeding age. By contrast, poor vision is quite rare in hunter-gatherer societies, where books and computers are also rare. It has probably been rare in human societies for most of our history.

But so typically for this tool-happy animal, our brains can cheat biology. The disk floating on my cornea is old technology. My aunt has just traded the entire lens of her eyeball for a plastic replacement. A friend recently had his eyeballs snipped to shreds with a laser, restoring their focus to childhood acuity. And scientists are developing an implantable light detector that can feed coarse data to the brain of a totally blind human. The owl of my childhood was able to detect a gnat at fifty paces, but if his vision had failed, he would have, too. When the human senses fail, whether from old age or too much reading, we invent another tool and look ahead to the next challenge.

EARS

Reading last evening on the couch I was distracted by a scratch. A scuttle. The plastery gritching of *Mus musculus*, the house mouse. My eyes leapt from the page and fixed themselves on a precise and particular spot on the ceiling. Nothing unusual appeared, but my eyes stuck there. My ears commanded them to. As a species with mediocre vision, I compensate by having really quite excellent hearing. Ha! The memory of my one-eyed father visits me again. He, driven half mad by the cavorting of squirrels, once discharged a shotgun into a precise and particular spot on the ceiling of his house. Perhaps his diminished vision sharpened his hearing, and his aim, but I'm afraid I don't recall the outcome of this particular experiment.

The point is, human ears are so sharp because our eyeballs show us only a segment of the world. By contrast, a rabbit, whose eyes take in a huge field of vision, has ears that strain to locate the ominous crack of a stick. In both animals, the two senses collaborate to ferret out the things that make life better or worse: food and friends, foes and falling rocks. And the two are so interlinked that when they gather conflicting information, the brain sputters nonsense.

The Pinnal Flap

So let's take a look at this pinnal flap of mine. It's no thing of beauty. It's a bald ruffle of cartilage, immobile as an owl's eye. Well, the corrugations have evolved not to attract mates, but to bounce sound waves into my ear canal for processing. These bounces help sort out the direction a noise arrives from. The shell shape bends slightly forward, but it's also open to sounds from the side. Its vulnerability is to the rear. Just as my predator's eyes leave me blind behind, my ears leave me with a "deaf spot." Humans have trouble locating noises that are directly fore, aft, or above our heads. And unlike a deer or a wolf, I can't swivel my pinnae for a closer listen. That said, a classic experiment found that humans instinctively waggle our heads like lizards when we're stumped, to get further hints from our pinnae.

Ear styles of the animal kingdom are almost as colorful as eye styles. My primate kin have equipment similar to mine—a bit smaller in the gorilla, sometimes larger in the chimp, but anchored on the skull, passively funneling data to the brain. Part of the panda's humanlike charm stems from the fact that its ears are almost stationary, fixed in an attitude of attentiveness. More immobile than the panda's are the ears of the true seals (harbor seal, gray seal, harp seal). Unlike eared seals (fur seal, sea lion), they have no pinnae at all, just fur-covered holes behind the eyes. Bird ears, too, are set within the skull.

A deer's ear, on the other hand, can swivel with the best of them. The front of the ear acts like a satellite dish, but the back is a shield, fending off distracting sounds. The net effect can boost the target noise by twenty or thirty decibels over the background hum. The best I can do in the deer's situation is cup my hands behind my ears and "shush" the humans around me.

As for ear location, I'm pretty normal. Most mammals separate their ears to maximize their triangulating potential. Still, they keep them handy to the brain that processes sound data. Exceptions do abound among insects, who may sport ears on their legs or wings. These oddly placed organs make more sense when you consider that the owners are sensitive to both airborne sound waves and earthly vibrations that arrive through their feet. Some moths even have ears on the abdomen or back. And while some moth "ears" may comprise a solitary cell, those paired cells suffice to determine the proximity of a hunting bat and to trigger an instinctive dive for the ground. If two cells are all you need to perform a task that complicated, you have an efficient piece of equipment.

Many reptiles and birds follow the mammalian model, placing an ear on each side of the skull. But to compensate for a narrow head, some have added a feature to help them locate the source of sound. My own ears are so far apart that a sound wave registers first in one, then a split second later in the other. The sound level drops, too, in the time it takes to reach the second ear. Between the time lag and the pinnae's input, my brain can calculate where to point my eyes. This triangulation doesn't work so well in tiny skulls. Thus birds and many reptiles have evolved a pipe connecting one ear to the other. Technically, I, too, have a tunnel through my head: Each ear's Eustachian tube opens into my mouth, so that I can pop both ears by closing my mouth and nose, and exhaling. But in birds, the connection is more open and free flowing. A sound passing through the right eardrum rushes through the tunnel and bounces off the interior side of the left eardrum. This provides an animal with the directional data it needs.

Frogs also follow this pattern but have elaborated further. Their inner ears connect to the lungs. In some frogs, those airbags are even more sensitive than the ears; in others, the lungs are sensitive to bass notes while the eardrums handle the treble. The lungs may also protect a frog's ears from his own bellowing by applying a healthy back pressure to the eardrum. At least one species, the Panamanian golden frog, has dispensed altogether with outer and middle ears, and relies on the lungs to inform the inner ear of the world's vibrations.

Wowl and his fellow owls boast notable ears, as well as eyes. A few species have remodeled their entire skulls in order to position one ear lower than the other. They enjoy binaural hearing that not only reveals the direction of a sound, but also allows them to gauge precisely the altitude of their target. Many owls also have a disk of feathers around each eye, rather like a satellite dish, which funnels sound into an ear slit just to the side of each eye. Pretty slick. (The "ears" that many woodland owls display aren't ears at all, but excitable tufts of feathers, like a dog's hackles. They may serve both to alter the animal's silhouette into something branchlike and to communicate silently with family members. Wowl would squawk and erect his plumicornes when he laid eyes on a broomstick, screwdriver, or any other kind of snake—remember, an owl's huge eyes leave little room for the brain.)

Acuity: Broadband Reception and Top-Shelf Triangulation

Whether an ear is located on the knee or the head, whether it's built of cartilage or lung tissue, its primary purpose is the same: to capture those particular vibrations most crucial to its host. Thus each species attends to a different subset of the sound spectrum. Sounds too deep or too high-pitched for its purposes pass by unnoticed.

A word about sound: Noise is a disturbance among air molecules. Some force—perhaps a bird squeezing air through its throat—sets air molecules banging into one another and rippling outward. The more vigor the bird uses, the stronger the ripples in the air: He gets louder. The higher his tone is, the more closely the waves are packed together. The *pip* of a chickadee launches a ripple of tight, soprano waves, while a raven's bass *croak* starts a ripple of long, lazy waves.

I, as an animal with boundless interests and behaviors, am interested in monitoring a huge variety of noises. The wind warns of an approaching storm. A grumbling in the underbrush alerts me to the presence of a large animal, who could be either predator or prey. My perception of bass notes is nearly as good as that of elephants and whales, animals whose telegrams are pitched low to travel through the soil and water, respectively. I'm not so good in the squeaky registers, though. My hearing fades out well below the vocal range of many small rodents, including mice, who, researchers recently discovered, warble complex love songs in a voice pitched so high a human can eavesdrop only through machinery. Owls, whose welfare depends more heavily on mice than mine does, can hear their love songs just fine.

Perceiving noise is one thing—any monkey can do that. It's how a creature reacts to information that advances its health and happiness. When it comes to using sound to guide my eyes, I'm unbeatable. When a mouse scrabbles in the wall or a door latch clicks in the house, my brain processes the data, then reflexively turns my head to point my eyes at the source. Most animals do this. But most of them can steer the eyes only to the general neighborhood of the noise. The eyes then take over the hunt. Not I. My brain can use sound to guide my eyes to within a degree or two of that noise. Time is saved; perhaps life is saved. This precision is shared by owls, bats, and huge animals like whales and elephants whose size enhances their triangulating. My fellow primates can't point their eyes any closer than three degrees from a sound source.

Not so impressive is my ability to detect a quick series of sounds. My ears, like my eyes, tend to process one noise, recharge, then process another, all rather slowly. At least compared to a bird. In my clunky ear, the song of the North American hermit thrush translates to *To-whee-chee-chee-chee.* But replayed at quarter-speed, the simple song explodes into a thousand notes replete with trills and arpeggios. Back at real speed, my ear misses all the nuance. The ear of another hermit thrush, however, operates quickly enough to register every syllable.

Now, as you might expect, if not from the previous portions of this book then from personal experience, male and female ears hear things differently. That infamous inability of mated humans to comprehend each other has a basis in biology. One recent revelation: My mate processes the sound of male voices in one part of his brain and female voices in another. When my mate hears a male voice, the data travels from his pinnal flap to that part of the male brain used for imagining how something might look. But a female voice, like mine, which tends to be higher and richer in modulation, proceeds from his pinnal flap to brain regions used for deciphering complicated sounds, such as music. The consequence? Although the world's newspapers squealed that males are deaf to female voices, British psychiatrist Michael Hunter suspects the opposite: that the brain region that males use to analyze a female voice is better suited to extract information. But Hunter hasn't proven that yet. Nor has he peeked into the female brain to see what my organ is up to when a male is speaking. For now, I like to think that my voice is music to my mate's ears.

Scientists have long known that male and female brains process sound at different speeds, too. My female brain registers a noise more quickly than does my mate's. At least two factors contribute to my feminine efficiency: estrogen and my smaller head, which shortens the commute time for the signal. The female sex hormone apparently accelerates the brain's noise processors. Some studies indicate that a female's fluctuating hormone levels sharpen my hearing on those days when I'm most fertile. As with my vision, my swift sound processing may be a mere side effect, or it could be an evolved acuity that helps a fertile female optimize her mating options.

If my brain outraces my mate's in processing the ears' data, his outpaces mine in responding with a physical reaction—slapping a whiny mosquito, grabbing a ringing phone, whirling to fire his spear at a mastodon. But in all these contests, differences are measured in hundredths of a second, so it's not something we notice around the house. Around the house, we notice that he likes music loud and I like it softer. And we both notice the mouse.

I was surprised when I learned that my eyes and ears collaborate so closely, complementing each other. But I also discovered that when the two senses disagree, all hell breaks loose. The confusion is known as the McGurk effect. I located a test of McGurk on the Web, and was suitably wowed: First I watched video of a man saying "Da-Da, Da-Da." It seemed a bit off. The first syllable sounded like "Ga." And the mouth looked as though it were saying "Ga," not "Da." Then I closed my eyes and played it again. I heard with unmistakable clarity, "Ba-Ba, Ba-Ba." Freaky! My brain had perceived a flurry of conflicting data—one story from the ears, which heard "Ba-Ba," and a different story from the eyes, which saw "Ga-Ga"—and then completely fabricated an explanation: "Da-Da." (To torment your own brain, Google "McGurk effect.")

The McGurk effect is a reminder that the human organs of sense merely probe for data. The real action unfolds in the brain. A rabbit, by contrast, can transmit an order for action directly from its eye nerves to its body, bypassing the brain altogether. The soybean plant, for goodness' sake, needs neither eyes nor brain to twist its leaves toward the sun. But in a human, it's the brain that analyzes the information, juggles the conflicts, and concocts a story to make sense of the world.

In general, my hearing is a respectable instrument. It brings me a large portion of the acoustic world, turning a deaf ear only to the hugest bass singers and the serenades of small rodents. My ears may not move, but between their crinkled flaps and their ability to snap my head around, they deliver a precision of targeting unsurpassed in nature.

NOSE, TONGUE

Taste: Middling

Speaking of senses that mush together, consider taste and smell. Though beginner textbooks still present them as separate senses, researchers file them both under "chemical perception." My tongue is a very blunt instrument. It's handy for speech, and it can sometimes be called upon to extract a celery string from between my teeth. But for chemical analysis, it's marginally more discerning than a piece of litmus paper. It can sort a handful of chemical families. But its partner the nose can discern ten thousand discrete scents.

My own tongue may be more discerning than many. I'm a "supertaster." When I first stumbled across the term, I thought it indicated a highly trained palate. Hardly. It indicates a hardwired hatred of Brussels sprouts. Supertasters make up one-quarter of the population, and more of us are female than male. We boast as many as seventy times more taste buds than normal tasters. Our overabundant receptors feed reams of information to the brain, which all too often responds, "Yeow! Spit that out!"

When I unearthed a list of foods that supertasters tend to hate, a smile spread across my food-intake orifice. Dark chocolate. Black coffee. Green tea. Chili peppers. Tannic red wines. Hoppy beers. Cabbage, grapefruit, and Brussels sprouts. Wince-worthy, all! Supertasters also shine at gauging the fat content of foods. This is my absolute forte. Show me seventy-seven ice creams ranging from full fat to fat free, and with a lick of each I'll put them in order.

At the other extreme, I once knew a man who belonged to the opposite one-fourth of the tasting population, the nontasters. (The middle two-fourths represent normal tasters.) The term is misleading. Nontasters can discern the same flavors I can. But with fewer taste buds than normal, they're rarely bowled over by strong flavors. This man could eat a jar of hot salsa for breakfast and found burned toast perfectly palatable. Vinegar? Loved it. Turkish coffee? He took his with a hunk of dark chocolate.

Why such a variation in sensitivity persists in the human animal is an interesting question. Supertasters I can understand: Bitterness in a plant usually indicates that it has evolved poisonous defenses against deer, insects, and other browsers. Thus sensing bitterness at the first nibble would be a valuable adaptation. It would be especially profitable in a breeding-age female, because fetuses are extrasensitive to poisons. And it would be profitable in a sick animal, because in small doses many bitters are curative—the root of twining goldenseal fights infection, poppy sap soothes pain, and a snakeroot plant lowers blood pressure. But perhaps the *inability* to detect defensive chemicals in plants, or the acids of unripe fruits, can pay off, too. Perhaps in lean times, the least finicky eaters harvest the most calories. The animal world suggests so: Grazing animals, who would starve if they spurned bitter plants, exhibit a high tolerance for the taste. Carnivores, who can afford to be choosy about their veggies, are much quicker to reject bitter foods.

A note about taste buds: Way back when, I was taught that taste buds are segregated on the surface of the tongue: Sweet and salty to the fore; bitter to the aft; sour to port and starboard. So wrong! Although some areas of the tongue are particularly sensitive to certain tastes, each taste can make an impression anywhere on the tongue. This has evidently been known to all but textbook writers for over a century. I'm also a little shocked to learn that I have taste buds even on the roof of my mouth and on the lid that covers my trachea, and maybe even in my stomach. Females, I discover, generally have more taste buds than do males, even among normal tasters. And lastly, since my high school biology days, a fifth taste has joined the sweet-sour-salty-bitter gang: Umami is the "meaty" taste of protein, unmistakable in aged cheese.

Most mammals are vulnerable to the bitter plant poisons, and can taste them. They also share my ability to detect sweet, sour, salt, and umami. After all, their dietary challenges are similar to my own. We all need salt to regulate the fluid in our bodies. As for sweetness, those of us who include fruits in our diet can cram in more calories per minute if we choose those most sugary. Sourness warns the tongue away from concentrations of acid that may be corrosive to the system. And umami signals valuable protein.

But each animal's palate is tuned a bit differently. Or a lot differently: Lions and tigers and cats (oh my) can't taste sweetness (oh bummer). Their carnivorous lifestyle, devoid of fruit, seems to have let that sensitivity lapse. Some birds apparently can't tell sweet from sour, perhaps because their dependence on fruit goes beyond the calories. (The cardinals and goldfinches in my yard, for instance, rely on fruits' color chemicals to tint their plumage.)

Species also vary in their sensitivity. The human sense is middling. I have about ten thousand taste buds to my name. The catfishes, who hunt murky rivers where food is hard to see, blow me out of the water with one hundred thousand. Receptors pepper their entire bodies, head to tailfin, and especially on the whiskers. These animals can sample a dish long before they open their mouths to eat it. The canids, on the other hand, are pretty dull. A wolf or my dog might have two thousand taste buds—one-fifth my number. Pigeons and chickens have less than one hundred, which could explain why they'll try to eat cigarette butts. Snakes lack taste buds altogether and use their tongues as I use my nose. Insects are a special case where taste is concerned. Various of them test a meal's appeal with chemical sensors on their feet, antennae, or forelimbs, in addition to their odd, insecty mouths.

And from there, on the point of a bee's antenna, we might leap to the subject of smell.

Smell: Not So Hot

My nose knows a lot more than my tongue. That little nub, the leading wedge of my anatomy, may even have led me to my mate, according to recent experiments. That might explain why I, like many humans, treasure the smell of my mate, and spread his T-shirt on my pillow when he's away. I would know my stepchildren by smell, too. My boy is grassy and wild; his sister more musky and earthy. Even the dogs in my life, I hesitate to confess, smell distinct: Kuchen has a faint whiff of leather; his neighbor Ruby's cheeks smell strangely floral. They couldn't be more different. And how powerfully they—and we—must smell to one another, given the keenness of their own equipment!

In the animal universe, my nub is often derided as a dud. Recall that one-third of my brain's real estate fires up in support of the data my eyes gather. The nose, by contrast, lights up the equivalent of a dim street corner. And compared to my dog's schnozzle, mine can't sniff its way out of a wet paper bag. But in spite of my sniffer's diminished circumstance, researchers interested in pheromones suspect it has hidden talents.

The most obvious survival value of my nose lies in its ability to detect dangerous foods before they enter my mouth. The sour and the rotten, the maggoty and moldy, all release fumes that my nose vacuums in. My nose can also warn me of approaching wildfires or thunderstorms. That's the easy stuff.

Sniffing out the ideal mate, that's where sniffing gets sticky. For most other mammals, this is a cakewalk. That's because most other mammals have a vomeronasal organ (VNO). Embedded in the nasal cavity, the VNO is a patch of cells specially receptive to pheromones. A pheromone, strictly speaking, is a chemical emitted by one animal that elicits an involuntary response from another. A mouse, who has a formidable VNO, sniffs the face of another mouse to whiff up pheromones. Her VNO forwards the data to the brain, where some three hundred genes are dedicated to VNO interpretation. Based on which chemicals she detects, the mouse may realize the other mouse is her own pup and lick it, or discover that it's a dangerous male and attack it, or recognize a sister and begin shifting her own fertility cycle to match so they can share parenting in the future. Or, she might perceive that this mouse smells like the perfect mate.

When an amorous mammal—perhaps including me—sniffs out a mate, she seems to focus on her prospect's immune profile: What disease is he vulnerable to? What disease is he immune to? Whether she finds this information in pheromones or some other body odor isn't clear. Whatever the messenger, it enters the brain via the nose. If our inquiring animal decides her beau's immune system smells too similar to hers, that could mean he's a sibling—a turnoff. If the profile smells too foreign, that could mean his race evolved in a different ecosystem, leaving him ill-adapted to hers. But if there's a modest degree of overlap in the two animals' immune profiles, then their offspring ought to inherit an immune system that's diverse yet suited to local conditions. She'll give him a whirl. Experiments on mice show that if you disable a female's nose, her social skills founder. She may neglect her pups or allow an intruder to kill them. A male mouse stripped of his nose gear becomes blind to both the charms of females and the danger of other males.

So mice, not to mention cats, dogs, elephants, badgers, elk, ponies, lemmings, and rabbits, have VNOs that can sniff out very personal information. I, however, have just the remnants of a VNO, its wiring lost to evolution. Whereas our mouse devotes three hundred genes to operating her VNO, I retain just two; the rest of my vomeronasal organ genes have devolved to blather. Even my regular nose tissue, which many animals can deploy as a backup VNO, is genetically impoverished. A mouse's nose lining bears 1,000 types of scent receptors. Mine has 350. And while I, and all humans, surely must give off pheromones, they're proving elusive to science. If my odor is soliciting uncontrollable responses from my fellow humans, I'm not noticing it.

But I probably do send chemical messages, and they probably bear similar information to those of other animals. My messages just aren't irresistible. Silk moths and pigs demonstrate the power of a potent pheromone. When a male silk moth's antennae detect female pheromone, he turns in to the wind and beats his way to the source. He doesn't have to mull over this decision. His response is automatic, triggered by her chemistry. Likewise, a female pig in heat: When she snuffs up a snort of male pheromone, she stands rigid as a show dog and arches her back, ready to hook up, as the kids say. Hog breeders, to facilitate the process of artificial insemination, entrance their sows with a synthetic knockoff called Boarmate.

But my putative pheromones are missing. At this writing, just one chemical has been identified that meets the definition in humans. (This in spite of the fact that an Internet search for "human pheromones" delivers about one million results, mostly related to perfumes and other products marketed with a subtext I can sum up as Boarmate for humans.) The newly identified pheromone, the charter and founding member of this long-sought guild, is (take a deep breath) androstadienone. The human male leaks it in his sweat, spit, and semen. Although females aren't conscious of its odor, when this chemical goes up the nose, the female body involuntarily ramps up the cortisol, an energy-boosting stress chemical. Eureka! (You reek—ah!) Why male-smell elicits cortisol is a mystery for now.

The reality is that it would be surpassing strange if I *didn't* have pheromones, when so many other animals do. And a trickle of evidence keeps whispering that humans can subconsciously sniff out a lot of information from one another's bodies, and that these chemicals do color our perceptions of one another. Check out these nosy nuggets:

- Within hours after birth, human mothers and infants can recognize each other by scent alone.
- In blind sniff tests, human females favor the body odor of a male whose immune profile happens to complement their

own. These females are subconsciously sniffing out suitable mates. (Scientists don't make subjects sniff actual bodies, but T-shirts that have been worn for a couple of nights. Sometimes they use pads that have been worn under actual arms, soaking up actual body chemicals.)

- In similar tests, homosexual males prefer the odor of both homosexual males and heterosexual females over the smell of straight males.
- Some males can detect which underarm pad was worn by a female during a cheerful experience; some males and some females can tell if a pad was worn by a male who underwent a scare. (Neither sex can sniff out a scared female.)
- Humans mistake the two T-shirts of identical twins for one human's clothing, even if the twins don't share a home. (So do tracking dogs, for what it's worth.)
- A breeding-age female whose menstrual cycle is erratic will tend to become more regular if she shares a shelter with a male.
- Males and females who smell the underarm secretions of females (but not males) report a reduction in their depressive feelings; female sniffers also report a relaxing effect.
- Female siblings who live together will, like mice, rapidly synchronize their menstrual cycles in about half of cases studied. About one-third of close friends who spend lots of time together will synchronize, too. Why? The most popular explanation is that, like mice, human females have evolved to share child-care duties, and this works best when the offspring are close in age. I have a hard time believing that a difference of a few weeks in infant age would be detrimental enough to drive the evolution of synchronicity. According to a less tender theory, fertility shifting is a war in which females try to move their fertile periods apart, so that they might win the undivided attention of what the researcher terms "high-quality males." Because too much shifting reduces mating opportunities, the

> cycle stabilizes. But if the motive were strictly competitive, then I would expect a "beta" female to actually match cycles with the "alpha" female, rather than shift away from her. She could then exploit high-quality males attracted by the alpha.

This wobbly list of clues (some findings rest on a single study) suggests that I'm oozing chemistry that can influence another human's behavior. But which of my odors and vapors are the magic ones? I have many smells up my sleeve. Scientists used to think humans lacked scent glands. But now they recognize that my armpits probably play an important role. These secret folds of skin produce a bouquet of chemicals, and bacteria living there alter them into a bonanza of additional chemicals. My underarm fur, if I didn't dispatch it with a razor, would provide the chemicals more surface area to spread across. And the heat of the folded skin certainly does cause the chemicals to vaporize into the air, on those occasions when I neglect to apply products that transform my natural scent to "flower fresh." My armpit odor, with additional chemicals seeping from other parts of my body, amount to what scientists call my "odorprint."

The research on how my odorprint may have affected my mate choices is far too young to answer such pesky questions as: Does my deodorant give me a deceptive odorprint? And, Does my perfume doom me to bad matches? These are questions for the next era in mate-sniffing investigations. Perhaps the most alarming research effort to date concluded that, of all the odors that can cause sexual arousal in a human male, the most titillating is that of cinnamon buns. Equally unnerving, females get hot and bothered over the mixed scents of Good & Plenty candy and banana bread. But due to the poor design of this study, I would not bet my genetic legacy on these attractants.

What we all want to know is: Did I or did I not sniff out my mate? I will say that my own mate's personal smell—the warm musk that scents his T-shirts and his pillow—is my favorite smell in the world.

But humans, burdened with fussbudget brains, base our mating decisions on a hundred factors, from height to weight, from income to an interest in beetle collecting. What percentage of my positive response to my mate was based on my nose? Who knows?

As I mentioned, humans aren't world-class chemical detectors. My tongue sorts through a mere five tastes: sweet, sour, salty, bitter, umami. It's my nose and the brain behind it that add flavor, the bouquet that blossoms in a spoonful of lobster stew or a slurp of hot chocolate. Even then, I'm limited.

My nasal tissue is equipped with receptors tuned to some 350 specific odors. When the bouquet of warm chocolate enters my nose, one receptor may fire or many; my brain receives their data and decides what I'm smelling. Thus my 350 receptors allow my brain to recognize ten thousand different chemicals. But give me more than four chemicals at once, more than four flavors in a mist of chocolate, and all but four will slide through my nose undetected.

By contrast, that mouse in my ceiling has 1,000 odor receptors to my 350. So a mouse, in addition to sniffing out personal information with her VNO, moves in a world of chemistry unknown to me. She can probably smell me through the plaster. And then there is my dog. There's some scent hound in his mix, and some of those nosy breeds devote twenty square inches of tissue inside the nose to catching molecules. I, by contrast, have two square inches. Kuchen also dedicates a far denser network of nerves to servicing that tissue. As a result, dogs can smell an odor far more dilute than I can, ranging from hundreds of thousands of times fainter in some breeds to many millions among the nosiest hounds. Birds, on the other hand, stink at sniffing. And our fellow primates are pretty dull, too.

As with taste, some humans have sharper sniffers than others. But smell is a notoriously unstable sense: My own ability, while fairly acute compared to my mate's, changes from day to day, and from one odor to another. Generally, scientists reserve the term "supersmellers" for humans who are hypersensitive to a single chemical. It turns out that some of us can consciously detect the male pheromone androstadienone, but we're otherwise normal in our sniffing ability. Ditto for asparagus pee: Some of us can smell that peculiar odor after eating asparagus; others lack the receptor for it. Professional wine and tea tasters are sometimes called supersmellers because they notice nuances in flavor. Perfume makers also employ expert "noses." But scientists agree that those humans aren't likely to have an inherent, genetic superpower. It's practice that allows them to make a fine-grained analysis of what's tickling their receptors.

As I've come to expect, *of course* males and females deal with odor differently. Many couples are already aware that females can detect odors at a lower concentration than males. But females are also quicker to identify those odors. Furthermore, female brains light up when they smell androstadienone, but male brains don't. On the other hand, male brains sparkle for the female hormone estratetraenol, which leaves female brains unmoved. And, perhaps adding a little evidence to the pheromone file, a female's sense of smell spikes sharply, like her hearing and vision, when she's fertile. Again, is this a hormonal side effect, or is the female nose particularly useful for judging male qualities at mating time? For what it's worth, both sexes of the ubiquitous European starling also amplify their sniff sensitivity during breeding season—by thirty times.

One fact of the smellable world still stumps me. I can understand my tongue's ability to sort out which foods are calorie-rich and which are toxic. I appreciate why my nose can sniff out rotten or ripe foods. I can even believe that I'm able to sniff out some personal data from the humans around me. But I'm stumped by my attraction to the smell of flowers. The local mayflower is a low woodland plant with a generic pink blossom. Because it depends on insects to spread its pollen, it has evolved a potent perfume that waltzes through springtime woods luring bees, flies, and me. I am not a pollinator. Flower blossoms are not a huge part of my diet. Why, then, am I so dazzled by the mayflower? I don't get it. No one else seems to, either.

TOUCH

Touch Reception and Processing

I've been privileged to pet lots of different animals, from mice to chipmunks, squirrels, rabbits, camels, and even a hedgehog. With the softest animals, a human goes through a sort of triage process to reap the full sensation of silkiness. We always start our exploration with the hands, of course. These are dense with touch receptors, especially at the ends of the fingers. But some animals, or parts of animals, are so soft that you can't even feel them with the fingers. Among larger animals, the flare of a horse's nose is in this category. In that case, if the animal is amenable, a soft-seeking human will apply a cheek. Here the receptors yield finer resolution. Here you can wallow in the velvet of that warm nose. A chipmunk, too, will register on the cheek when the fingers are too blunt. (Applying cheek to chipmunk can be a challenge. It takes days and many pounds of sunflower seeds. But it's worth it.) And then there are some furs so fine we must turn the cheek and apply the lips. The human lips are jam-packed with sensors, and they are the only organ that can do justice to a flying squirrel. That creature simply disappears in human hands, too silken to leave any impression beyond warmth. Usually, of course, it's Kuchen I turn to for a fur fix. It's a reliably positive experience to bury my hands and face in his glossy, brindled fur.

Every creature experiences the physical world in a slightly different way. The naked mole rat, a Kenyan earth dweller, does most of his feeling with a set of outsized incisors. The regular lab rat (aka the Norway rat) explores her night world through long, sensitive whiskers. The raccoon reconnoiters with sensitive paws. The platypus navigates with a broad, rubbery bill. These biases are mapped in each animal's brain. When scientists chart which part of the brain lights up in response to a given touch stimulus, they discover quite a different map for each animal's body. The mole rat, for instance, devotes 30 percent of its "body map" to four front teeth, short-changing its other body parts in the process. The Norway rat dedicates similar acreage to its whiskers. In a raccoon's body map, the forepaws monopolize half of all the available territory. And the platypus's bill is even more autocratic, sparing just 15 percent of the brain map for the rest of the body. Drawn to proportion, the platy map looks like a banana of a bill with a Brazil-nut body tacked on.

And my map is just what you'd expect from a squirrel kisser. I've long wanted an opportunity to use the word "homunculus." Meaning "little human," it's a sketch of the human body in proportion to how much brain it uses to process touch information. The human homunculus is nearly as freakish as the platypus's: The upper half of my skull is a baseball; the lower half is a basketball, a great balloon of cheeks, mouth, and lips. A twig of a neck sprouts twig arms, which balloon into catcher's-mitt palms with baseball-bat digits. My torso and legs are twigs, then clown feet finish the portrait. That's how my brain allocates space to my body's physical experience: the lower face, plus hands and feet, dominate.

It is also biologically reasonable that I stroke my dog, versus laying my hand still against him. The perception of velvetiness, like the perception of prickliness, stickiness, and scratchiness, depends on motion. If the texture sensors in my skin stagnate, they doze off. I experiment with this, stilling my hand on the ruff of Kuchen's neck. Heat I feel, coming up from the skin. Pressure I feel, too. But the sensation of softness fades in a split second.

Touch isn't a single sense. It has many components. Scattered through my skin, special sensors with fabulous names—Merkel's disks, Krause's end bulbs, Ruffini endings, nociceptors, and corpuscles of Pacini and Meissner—relay separate categories of information to my brain. I place my hand on Kuchen's ruff again. Under my skin, the supersensitive Pacinian corpuscles are quick to fire off data on vibrations as the fur shifts under my fingers. But Pacinis tire quickly, and my experience of texture vanishes. Some Meissner's corpuscles closer to the surface have fired, too, giving a reading of pressure and a rougher reading of texture. Although they have more stamina than the Pacinis, they weary of signaling after about a second. My good old Merkel's disks display more staying power. "There's pressure," they report to my brain. "There's pressure. There's still pressure." They may persist for a minute or more. Simultaneously, Krause's end bulbs may note a change in temperature, as heat from my dog meets my skin. These temperature receptors are steady workers who would quit only if my dog caught fire and the temperature against my skin reached 113°F (45°C). At that point, naked little nerve endings called nociceptors ("harm takers") would snap to attention and phone in a pain alarm. Those nociceptors would also call in an alarm if my dog froze beneath my hand. At least that's the theory. The sense of touch is still something of an unexplored territory.

Much of this system is shared by my fellow vertebrates (animals with backbones), although we all distribute our receptors according to lifestyle. Herons and cranes, for instance, pepper their legs with Pacini-like receptors in order to feel the tickle of prey moving through the water. Ducks, sandpipers, and other birds who dabble in water wear their Pacinis on their bills, for the same purpose. Bats sport mountains of Merkel cells on the surface of their wings to sense air turbulence, allowing them to fly by Braille. And, of course, my dog's own Merkel's and Pacini-esque sensors ripple off messages to his brain as my hand smoothes his head again and again.

Importance to Normal Development

Patting my dog is more than pleasant, it's healthful. All those cells and bulbs and disks need stimulation, especially in their youth. If a young animal is not touched enough, things can go wrong. Had my parents neglected to pat me, I might now be a jittery, anxious wreck. Such sad effects are best demonstrated on rodents, whose simple lives limit the variables that can complicate an experiment. Also, their parents are less likely to object. The results: Study after study has shown that rat and mouse pups who don't get much licking from their moms grow up with maladjusted emotional systems. They're more fearful and stressed; they're binge eaters with subpar development of their brain cells. Touch-deprived females make particularly screwy and neglectful mothers whose innate maternal behaviors fail to activate.

Human babies also suffer if they're undertouched. Premature infants are most at risk because they need so much medical intervention that parental cuddling falls by the wayside. This research pool is a bit turbid, however. Many studies show that "kangaroo care," the snuggling of a naked preemie against a parent's skin, produces more emotionally stable babies who enjoy lower stress levels, stronger appetites, and may leave the hospital earlier than touch-deprived infants. But variables preclude a clear view: Are the babies eating more because they're nursing upright, versus prone? Are they less stressed because a parent's heat is more stable than the heat inside an incubator? Are they responding to a parent's touch or to motion? Rat research suggests that mechanical "patting" will produce a well-adjusted rat orphan. A cynic might suggest that the biological basics of mice and men are similar enough that what's good for the rat pup is good for the kid, and prescribe patting machines for babies. Suddenly, I recall a childhood penchant for napping atop the clothes washer. I'd clamber up, bunch up a towel, and drift off to the warm chug-chug-chug of the machine. If I am a particularly well-adjusted human, perhaps it's due to the attentions of my mechanical mommy.

Perhaps the touch response of adult humans can help clarify the importance of physical stimulation. Most of the research on adults is conducted in hospitals and involves massage and therapeutic touch. And there, at least, the power of touch isn't working miracles. Most studies agree that touch doesn't reduce the need for pain medication, in spite of some patients claiming it reduces their pain. Nor does it relieve depression. And it's not a sleeping pill. Touch does, however, relieve anxiety. Humans do seem to go more gently under that steel knife after they've been massaged or otherwise touched for twenty minutes or more. If touch can reduce anxiety in adults, it may also reduce anxiety in infants, which would certainly be worthwhile.

Pain Threshold: Variable

Speaking of pain: As I sit, so do I ache. The right side of my neck and shoulder are wracked by muscle spasms, left over from a car accident years ago in which the momentum of my body, which evolved to change speeds rather gradually, was sharply arrested. The flower stalk of my neck was torn between following the cabbage that is my head and staying with the rest of my body. The left side throbs with its own spasms, acquired more recently when a large vehicle attempted to mount my little Earth saver, dunking the Earth saver's roof onto my head and crunching the flower stalk into my left shoulder. It's popular to complain that the human animal did not evolve to sit before a computer for hours at a time, and that this causes all our problems. But the fact is that in computer-free cultures females also spend hours sitting and working with their hands. And as their bodies accumulate the years, they, too, gather torn muscles and knots of pain. As do the bodies of lionesses who tackle zebras for a living, and moose bulls who butt heads over breeding rights, and squirrels who fall out of trees (it happens more often than you'd think). Pain happens. When heeded in time, pain warns an animal to slow down, back off, run away before it's too late. At worst, it settles in for life. And so as I sit, I ache. I like to think I don't whine as much as my mate, who's peppered with arthritis and skiing knots.

Recently research has been generous in its treatment of the spleeny. Where once humans were divided into just two camps—stoics and whiners—a Michigan neuroscientist now grants us three: val-vals, met-mets, and val-mets. (The mumbo jumbo stands for variants of a gene.) Each category represents a different combination of a mother and father's *COMT* gene. This gene comes in two varieties: good painkiller and not-so-good painkiller. One-quarter of humans are val-vals, who got the painkiller version from each parent. Half of us are val-mets, with moderate pain control. And the final quarter are met-mets, with two lousy versions of the gene, which essentially allows pain to run roughshod through their brains. So if you whimper over splinters, you're not a sissy, you've just been cheated by the Gene Fairy. Other mammals share the *COMT* gene with us, and research implies that they, too, come in strains of Spartan and simpering.

If I do simper more than my mate, well, the male and female human perceive pain differently—even after accounting for the *COMT* gene. The sex hormones that soak our respective brains seem to dictate how each sex will experience life's cuts and contusions. Generally, females have a lower threshold: It takes less burning, freezing, poking, or pinching to make them yelp. They also have a lower *tolerance* for pain: They're quicker to capitulate and pull their hand from a bucket of ice water or a heat beam. (These are common tools of torture for pain researchers.)

Other differences hint at the separate ways that male and female bodies react to pain. During painful times, a male's heart beats faster but not the female's. A male's cortisol and endorphins (stress chemical and painkiller) rise, but for some reason, a female brain neither stresses out nor self-medicates for pain. For a female, the sum of three pinches hurts more than the individual parts, while this is not true for males. This overall sensitivity to pain holds true in lab animals, too—female rats, like humans, are also more vulnerable to the alarms of their nociceptors. Why? Most theories cluster around a female's role in reproduction: She must avoid harm "for two," so to speak.

Curiously, the human female's pain sensitivity goes dull when her estrogen level rises toward the fertile days of her monthly cycle. This contrasts with the pattern among the other senses, all of which sharpen for the peak breeding days. Whether a dulled sense of pain aids a female in her conquest of the ideal gene donor, it's impossible to say. It could—as could the rest of the pattern—be a mere side effect, not an evolved effect.

For a sense that engages so much of the human body, extending deep into the interior, science has paid surprisingly scant attention to touch. That we still can't say exactly how a Ruffini ending facilitates a human's perception of the world is peculiar. But perhaps it's true that touch is the least critical of the five. With the chemical senses, we test the nutritious and noxious elements that might enter our bodies. Vision and hearing are early warning systems that report the approach of both hazards and opportunities. But touch, well, perhaps the information it provided was too little too late for our evolving ancestors. If a hazard or opportunity had already approached so closely that it was rousing your Ruffinis, maybe it was too late to gather much more useful data. And maybe that "second-string" status makes touch uninteresting to scholars of human perception even today.

BONUS SENSES

That concludes the census of the human senses. Or does it? So many other features turn up in the animals around us—senses of direction, of electrical fields, of tiny changes in temperature. Am I really numb to those things? Or are my extra senses lying dormant?

Direction

A sense of direction seems most likely in humans because we do have shards of a compass scattered about the brain. Magnetite is a naturally magnetic form of iron. Organisms as diverse as bacteria and rainbow trout string together magnetite crystals in their brains. The chains then (presumably) pick up the same geomagnetic fields as a compass needle, helping the organism to chart its course around the world. Magnetite is implicated in the navigation of salmon, honeybees, loggerhead turtles, pigeons, mosquitoes, newts, and mole rats, among others. The degree to which each animal relies on magnetite seems to vary. When scientists outfit a pigeon with a device to foul magnetic fields, the bird sometimes loses her ability to navigate. But salmon, treated to the same handicap, carry on without discernible difficulty. It's all a bit fuzzy yet, even in species other than my own.

It is confirmed that magnetite crystals dwell in my own brain. But whereas a salmon's crystals concentrate in the nose, and a pigeon's cluster at the base of the skull, mine lack a rallying point. They're scattered. This is not what you'd expect of an efficient perception organ. So it's possible that humans once did boast a functioning magnetite compass, but it has fallen apart from disuse. It's also possible that the crystals are an accidental by-product of some other process, or that they do some task beyond my power to imagine. Regardless, a human who has lost her bearings would be better advised to rely on the north/south biases of plants and the location of stars in the sky than on magnetite chains in her brain.

Or better still, she might ask a male. The average male brain can more easily perform "mental rotation" of an object. That factoid seems a bit arcane. But perhaps it relates to the different ways in which male and female humans throughout prehistory exploited the resources within their territory. Judging from modern hunter-gatherers, females typically circulate among patches of stationary food: a fruit tree here, a patch of tubers there, a nut tree over the hill. Males certainly forage, but they specialize in hunting. And prey animals, in contrast to nut trees and tubers, run around. One result is that females may travel just a couple of miles a day, and males may travel ten times as far. Some researchers speculate that the space-savvy male brain evolved because all the hunters with bad spatial skills got lost and died childless. Modern males are believed to navigate by building a mental map of where they've been and which direction they're headed now. And the female brain, according to recent experiments, prefers to navigate by landmarks: Take a right at the light, left on Main, left again at the tuber patch. Since hunting flighty animals while carrying a fussing infant was a fool's errand, those females who could recall the location of plant foods evolved a brain that shines in the task of "object location." To this day, the typical female can easily locate a jar of mayonnaise in a crowded refrigerator and guide her mate to this object—even from her office twenty miles away.

Electricity

How about electrical perception, then? Some sharks, rays, and electric fishes can sense the electrical field of a nearby prey animal or even send an electric telegram to one of their peers. Me? Other than the occasional static-electric spark that passes between me and another animal, I can't claim much grace in navigating the electrical world. Human bodies are highly electrical—nerve impulses are all about electricity—but it's very weak stuff.

Heat

Temperature gauging I can do to some degree, thanks to the touch sensors in my skin. But I've got nothing like the equipment a pit viper deploys. These animals—the rattlesnake, bushmaster, copperhead, bamboo, and other venomous snakes of Asia and the Americas—have evolved small pits below the eyes that receive infrared wavelengths. This is the gentle heat that my body, and, more to the point, a mouse's body, radiates into the world. Hunting at night, a pit viper "sees" a mouse not in black and white, but as body heat against a cool background. With twin sensors, a viper even gets a "binocular" view. ("Bithermic" would probably be more apt.) The pit vipers can detect a small, warm body more than a foot away and can sense a temperature change of a fraction of a degree. It's a view of the world I can get only through technology.

Time

Is my sense of time a real sense? Like many humans, I have an uncanny ability to guess the time. This evaporates when I spend time outside of my clock-driven culture—camping in another ecosystem, for instance. But when I'm enmeshed in the minutes and hours of my daily schedule, I can usually gauge the time to within a few minutes.

As is often the case now, researchers learn about such mysteries by studying humans whose brains have been damaged. We now know that the human's inner clock resides in the basal ganglia (deep under the brain) and the right parietal cortex (above the ear). And indeed, a sense of time is something you can lose. Humans afflicted with Parkinson's disease and attention-deficit/hyperactivity disorder both lose their grasp on time, and can recover it with the aid of drugs.

Like other animals, I set my internal clock by the sun. Humans who are jet-lagged or deprived of light stagger around the clock for a while before they settle into a new rhythm. My dog has similar difficulties when daylight savings time arbitrarily delays his dinner hour. He needs a week or two to adjust. It's the long view of time, the concept of a future, that sets a human apart. When a squirrel buries a nut, it's a near certainty she's not conscious of her future. She's mindlessly repeating an instinctive behavior she inherited from forebears. But in a handful of species besides mine, scientists have discovered a modest capacity for planning. Orangutans have shown they can plan at least as far away as tomorrow; ditto for a jaybird. (More on these in chapter 8.) But most animals take one day at a time, relying entirely on the timepiece in the sky that my ancestors evolved with.

Balance

What about a sense of balance? My Pilates instructor loves to torment the class by raising us on our toes then making us turn our heads left and right. The heart of the balance sense does have a home in my body, which I've forgotten all about since high school biology class. Crammed in with my hearing instruments is the vestibular apparatus, which resembles a squid eating itself. The business end of the squid contains chambers of fluid and little crystals that each register some component of motion—rotation, acceleration, deceleration. And yes, they can malfunction, ruining a human's chance at catching a wildebeest or anything else. The easiest way for a normal human to scramble her sense of balance is to go for a sail or read a book in the backseat of a moving car. The squid in her ear registers the bouncing, but the eyes beg to differ. Then, theory has it, the brain concludes the eyes must be hallucinating, deduces that the mouth must have gobbled something horrible, and directs the stomach to heave up the offending article. So, yes, balance is a secondary kind of sense, and it can get into snits with the eyes.

All things considered, how do my senses stack up? Thanks to my enormous brain, I do fairly well. I can dedicate one-third of the noodle to eyesight, but unlike an owl I still have room to support a decent intellect.

Visually, I take a predator's view of the world. My eyes are oriented for pursuit and capture, at the expense of protecting my back. My ears help to compensate for that. My chemical senses also suggest a predator's lifestyle: My most sensitive taste is for bitterness, a sign that I can afford to be picky about which plants I eat. Inside my skull where all my sensory data are crunched, the processing times aren't what they could be—or what they are in a housefly, for that matter. But in part that's the downside of the big brain: It takes time for information to travel across that great gray nation. And, yes, as with other animals, it's a scientific fact: Males and females see—and hear, taste, touch, and smell—two different worlds. But perhaps we had already perceived that.

4

FREE AS A BIRD:

RANGE

Homo sapiens inhabits a vast range that runs from the equator to the Arctic. Thanks to an omnivorous diet and an ability to make shelter from many natural materials, this species has been able to expand far from its evolutionary birthplace in Africa over the brief span of sixty thousand to one hundred thousand years. Some of the predecessors of *H. sapiens*—*H. erectus* and *H. neanderthalensis*—displayed this same flexibility of habitat and propensity for range expansion.

Although the races of humans in extreme ecosystems have evolved special features, these differences are subtle and can be difficult to recognize in the field. Among them: Humans who evolved in the coldest parts of the range tend to be taller and heavier with shortened limbs and narrow noses, thought to conserve body heat. Contrarily, humans who evolved in the warmest habitats are inclined toward small size and/or elongated limbs that efficiently shed heat. Quite a different set of adaptations appear in those humans evolved to exploit high-altitude ecosystems, these involving modifications to the lungs, heart, and blood. And humans who evolved in arid zones, while less studied, appear to manage their fluid balance more efficiently than other races. Furthermore, all humans possess such a flexible biology that they can acclimate temporarily to any ecosystem in the animal's entire range.

This primate's remarkable distribution is matched only by that of a tiny number of land animals. Even migratory birds who summer and winter in habitats thousands of miles apart can't survive a full year in either ecosystem. The closest contenders are the domestic dog and species like head and body lice, whose natural history dictates that home is where the *Homo* is.

NEAR-GLOBAL DISTRIBUTION

I once spent a spring day among the Nenets reindeer herders of northwestern Russia. I did not know how much snow was on the ground, and when I jumped from the helicopter I sank to my crotch. I labored toward the Nenets camp by raising one leg from its hole, swinging it forward, falling onto it with chin in the snow, then dragging forth the other leg. It was more like swimming than walking. The brown faces of my waiting hosts crinkled in amusement. Their blue canvas tents stood near a clump of spruce trees where the sun had warmed the trunks and begun to reveal circles of earth. Carved wooden skis, with a strip of caribou fur on the kick, rested against a reindeer sleigh. Pausing to examine the sleigh, I could find no evidence that the Iron Age had passed this way. The runners were bent saplings. The decking was more of the same. All were joined with wooden pegs and lashed tight with strips of caribou hide.

We were dropped here, a group of journalists and innocent civilians, as part of a test run for a commercial tour near Arkhangelsk on the White Sea. The Dutch organizers hoped to convert the local economy from clubbing baby seals to photographing them. For the most part, the tour was a bust, with our two Russian guides (whose personal economies had been forcibly diverted from the engineering sector) feeding us canned peas as we huddled in the frosty barracks of the seal clubbers, our beds pushed against the door to repel the hospitality of drunken hunters. The baby seals had already been clubbed or had grown up and left; the engineers, mistaking us for nature haters, menaced and chased the adult seals over mushy, perilous ice floes. But then, as a diversion, we were choppered to the central district of nowhere and left for a day with a band of Nenets.

When I staggered onto firm snow, I was greeted by Ilya, a slight male with a shock of black hair and a huge smile. Behind him was his family's tent. The tents of a few other families clustered nearby. About one hundred of the group's tamest caribou milled, hooves clicking, in a sapling paddock. The clothesline in Ilya's dooryard was hung with drying caribou hides—raw materials for most of what he required. And Ilya himself was suited entirely in caribou hide—back fur on his back, leg fur on his legs. He was the very picture of a northern human.

I've had the opportunity to visit many extremes of the human range. In the course of this chapter I'm going to revisit them, starting in the dry, cold latitudes, and ending at the hot, wet equator. On my initial visits to these ecosystems, my own physical adaptations proved to be temporary and modest. And sometimes painfully slow. But in each spot, I encountered human races whose long history in that habitat had resulted in permanent variations on the human theme. Long before the invention of pith helmets and GORE-TEX, the human animal spread to most of the habitats on Earth. The cold tundras have been colonized. Arid stonescapes blasted by sun are no obstacle to human life. Humans even inhabit jungles where as much as forty feet (twelve meters) of rain fall in a year. We see the dynamic in action in recent history: When farmers overfilled Europe a few centuries ago, the most adventurous of them struck out for fresh fields. Wherever they encountered natives that could be dislodged, they elbowed them aside. They discovered savannas in the Americas, Africa, and Australia, cold forest in my neck of the woods, and wet and dry jungles in Central and South America. And in each case, these newcomers mastered the challenges of the unfamiliar ecosystem and continued to reproduce.

This ability to conquer novel habitats predates *Homo sapiens*. An unknown pre-*sapiens* hominid hiked clear to latitude 40° north about 1.5 million years ago, leaving stone tools but no helpful bones that would identify the hiker in question. That latitude equates to modern Chicago, or Kyrgyzstan, or Beijing. It's not Arctic but certainly not tropical, either. Although the planet back then was generally warmer, the climate also fluctuated wildly, leaving anthropologists to conclude that this mystery human had mastered life in the cold. Then, more recently, *Homo erectus* walked and yachted clear across southern Asia to the remote island of Flores about 700,000 years ago. This demonstrated an aptitude for hiking, and for shipbuilding and navigation, too. By 300,000 years ago, another mystery hominid was discarding stone tools way up in Siberia, a stone's throw from the Arctic Circle. And 200,000 years ago, *Homo neanderthalensis* nestled into the harsh tundras of glacial Europe, wolfing down mammoths and shrugging off wind chills estimated at –11°F (–24°C).

When it came time for *H. sapiens* to reenact the hominid expansion out of Africa, we were a fundamentally tropical animal. We bore negligible insulation from either fat or fur, and our metabolism burned just warm enough to stay comfortable in the tropics. Nonetheless, by fifty thousand years ago we had penetrated the chill of southern Australia, where we would cling even as climate change desertified the lakes and drove sand dunes across the continent. At about the same time yet another enigmatic member of the genus *Homo* shouldered north of the Arctic Circle in northern Russia. Temperatures there would have been about 18°F (10°C) colder than today, for winter lows of –45°F (–43°C). Once again these hominids left only tools and the bones of prey animals, so we can't know if they were *H. sapiens* or *H. neanderthalensis*. But clearly these *Homo*s could contend with endless winter nights and truly Arctic weather. And clearly hominids were determined to expand their range.

Range expansion happens. And not just to us. It's in the DNA of living things to secure themselves room to thrive, even if that means hiking a bit. A couple of years ago scientists unraveled genetic evidence of globe-trotting in a land snail, of all things. The species expanded out of northern Europe to the Azore Islands, then onward to the Atlantic isles of Tristan da Cunha 5,500 miles away—all presumably on the feet or feathers of birds. A few years before that, biologists were delighted to witness, firsthand, driftwood depositing a load of iguanas on the beach of a formerly iguana-free Caribbean island. And the European starling, once at home only in Europe, has lately colonized all but the coldest and driest habitats on Earth. Of course, for every success story, another species demonstrates a stubborn commitment to a tiny slice of the planet. Many living things are hog-tied by the specificity of their needs. Our dear relative the chimpanzee resembles the human in many ways, but an ability to live outside of warm forests is not one of them. As humans have chopped down the chimp's habitat, the entire species has concentrated itself in a shred of central African forest. When that's gone, the chimps are, too. A comparable fate confronts the other great apes, as well as the bamboo-bound panda bear and various salmon species whose spawning migrations have been cut off by dams.

But humans, when we have felt penned in, have always struck out for greener pastures, redder deserts, or bluer waters. My own ancestors reenacted this behavior just a few hundred years ago, fleeing Europe to cast themselves on the mercies of northeastern North America. Each time humans have moved from the tropics to the cold extremes, we have dreamed up new tools to buffer ourselves, or sometimes simply endured the discomfort until natural selection provided genes more suitable to the new range. In rather short order, our range covered the whole terrestrial world, excepting a few deserts, glaciers, and mountaintops. We settled below sea level (Holland) and three miles in the air (Tibet). We hiked and kayaked around the Arctic Circle and ate cactus pads in the Sonoran Desert (United States). The only creatures who gave us a run for our money were those who evolved to depend on us: the dog and a few parasites like the body louse who would instantly go extinct without us. We are the species with the greatest range on Earth. And again, we mastered all these ecosystems many millennia before the invention of Polarfleece or sunscreen.

COLD AND DRY HABITATS

So, back to the dry, cold north, where Ilya shows me around the business end of a reindeer. Humans who are comfortable with animals can spot one another, and without five words of common language, Ilya and I discuss harnesses and botflies and dogs. Farm girl that I am, I end up taking Ilya's reindeer sleigh for a spin. Compared to modern horse harnesses, the hide strips are thin and harsh, but the five deer trot forward steadily, tongues hanging out.

It's late winter, and these primates and ungulates are drifting north, out of the forest and onto the shrubby taiga. When the hundreds of reindeer now scattered across the land have scratched through the snow and eaten all they can, they'll amble onward. Then Ilya will pack his tent, fur blankets, and children, plus a few cooking utensils and a small stove, onto the sleigh. With a whoop to the bushy dogs, the whole civilization will creak and click toward better grazing. To the west, the Sami of Norway have an enchanting word for their nomadic lifestyle: To follow the caribou, aka reindeer, is *reindriften*.

I think of northern humans as dark-skinned, but Ilya's infant (clad in reindeer calf hides) is creamy. Her eyes show the epicanthic fold typical of the Mongoloid race. Her eyebrows are barely bumps in her skull, whereas mine cantilever over my eyes. Her face is semi-spherical like a ball, with a flat nose, minimal browridge, and cheek pads that swell gently between cheekbone and jaw. It makes me think of the round Mongolian *gers*, or yurts, designed to shed the wind that streams across the steppe. All the Nenets in this camp share these features, sometimes called the "cold-engineered face" by anthropologists.

For the moment, my attention shifts to my own adaptation. Humans, despite our tropical roots, are able to make temporary adjustments to all manner of extreme conditions. My adjustments are progressing a bit slowly. Like a small percentage of humans worldwide, I have a cold-protection system that overreacts. Every human's fingers get colder in the winter—by about 5°F (2.8°C). Mine drift lower, to 6.5°F (3.6°C) below normal. This is a sign of Raynaud's phenomenon, a genetic peculiarity that blossoms most readily in humans who have lived in a cold climate. It also turns up more often in females than males, who are generally harder to chill anyway. And it can be hazardous. When I was plowing through the snow to the Nenets camp, my working muscles radiated heat as a by-product, boosting my temperature. My fingers were pink inside my mittens. But when I stood still, lingering in the snapping-cold air, my core temperature dropped a click. And then, in a fit of cold-induced hysteria, the blood vessels in my fingers and toes squeezed shut, shunting more blood to my vital organs. My fingers turned white, joint by joint. I could have dipped them in the ptarmigan soup steaming on Ilya's stove and felt nothing. This vasoconstriction response would be natural, even desirable, if I were truly at risk of freezing. It prevents hot blood from cooling as it flows through the extremities. Although this invites frostbite of those extremities, it safeguards the more critical parts of a human. But just why a few percent of us jump the gun into this cold-protection mode is a mystery. I shove my hands deep in my down pockets and try to keep moving.

As for the rest of my cold-adaptation repertoire, I'm probably making small progress. I would need to spend weeks or months in the cold—hours a day outside—to maximize my potential. And even then, I'd be nowhere near as hardy as the Nenets. Temporary adaptation simply can't compete with eons of natural selection.

Some of the techniques Ilya and I use in a cold emergency are identical. One of the simplest modifications is that my sensation of pain dulls. The stimulus (steely, icy coldness) is the same, but the response (agony) relents a bit. Also rather quickly, normal hands will proceed to a "hunting reaction," alternately cutting off circulation to save heat, then opening the arteries again to prevent freezing. (My Raynaud's hands don't hunt.) If I dallied regularly here in the Arctic, my metabolism would get in a habit of dialing back a click when I got a chill. This would conserve fuel but leave my whole body 1.8°F (1°C) colder. Also to economize, my shivering response, designed to generate emergency heat by exercising the muscles, would kick in at a lower temperature. And along with my sense of pain, my perception of cold would fade. I might put on a little more fat.

Under truly dire conditions, by chilling myself severely, I could become a cold-weather beast in a few weeks. Scientists have soaked human males in tanks of 57°F (14°C) water for an hour at a time to get precise measurements of what happens. They found that cooling a male for an hour every other day, for four to six weeks, was enough to harden a body into cold condition. By the end of the experiment, the males were burning 20 percent less energy to keep themselves comfortable in the tank. And that's what acclimation is all about: making a more efficient animal.

These are neat tricks, especially for humans inhabiting a temperate environment like mine. We animals of the mid-latitudes have to grapple with four distinct seasons, from hot to wet to cold and back to wet again. In fact, my response to cold, and that of other temperate-evolved humans, may be more robust than that of dark-skinned humans from warmer climes. Black American soldiers, who inherited their African ancestors' gift for heat resistance, are four times more vulnerable to frostbite than I am. I say my response "may" be more robust than that of dark-skinned humans because I can find studies that deal only with populations of African ancestry. This neglects dark-skinned populations like Melanesians and South Indians. However, the fact that spotted guinea pigs get frostbite on their black spots before their white ones suggests it is the skin-tinting melanin that makes the difference. If that's the case, then any human with dark skin will face a higher frostbite risk than a pink-skinned specimen like me.

Proud as I am of my short-term acclimation, Ilya has me cold. Northern natives have access to all the same emergency measures I have, but they can take them to extremes.

How Ilya himself, and his relatives, would combat a super-cold night, if caught far from his stove and caribou blankets, science hasn't determined. Only a few cold-realm populations have been analyzed. But judging from those, Ilya has evolved one of three popular approaches to weathering a blast of cold. I'd bet he does what his reindriftin' neighbors in Scandinavia do. They actually call on two of the three techniques at once: insulation and hypothermia. If this is the case, when Ilya gets a chill, his circulatory system will cheat the skin in order to insulate his blood from the cold. His skin will grow even colder than mine could safely do at my most acclimated. At the same time his metabolism will drop, cooling his vital organs but saving fuel. Again, his metabolism can drop lower than mine without hazard. This two-pronged defense, called "insulative hypothermia," is usually the response evolved by cold-realm humans who don't have access to boatloads of rich food.

It was not Arctic humans but the Aborigines of central Australia who made insulative hypothermia famous, because they can do it virtually naked. Come the cold desert night, they toss together their notion of shelter: a screen of branches to break the wind. They let the fire die and bed down in their birthday suits. Down, down, down slide their core temperatures, falling 7°F (4°C) by sunrise. They sleep like babies. When sadistic scientists outfitted Caucasians for the same lifestyle, the subjects spent a night shivering and sleepless, their metabolisms spiking wildly up and down.

If Ilya does not practice insulative hypothermia, then he might do just one or the other, either treating his skin as insulation, or chilling his entire body. The Aborigines of northern Australia, where food is more plentiful than in the desert, employ only the insulative response: In a cold snap, their skin temperatures dip, but their metabolisms and core temperatures hold steady. In contrast, the strictly hypothermic path is taken by both the Bushmen of the Kalahari Desert and by Peruvian Indians in the Andes mountains. Rather than singling out the skin for sacrifice, these humans let the whole body cool to conserve calories.

And finally, it's possible that Ilya would use the third option: raising his metabolic rate, instead of letting it languish. Among humans with a ready supply of food, the body often opts to stoke its fires and let a raging metabolism warm the corpus. The Inuit enjoy a high-animal diet that's swimming in fat. Inuit don't store this fat on the body; they burn it right away. This comes as a surprise to me. I always thought the Inuit had well-padded bodies, and I guess I've never had an Inuit friend with whom I might disrobe to compare morphology. I've encountered mainly Inuit faces, and they lend an impression of plumpness. But now, having come across some photos of Inuit males in their skivvies, I stand corrected. To grossly generalize, they're as lean as anyone else and more muscular than most. Scientific "skinfold thickness" measurements bear this out: Inuit carry no more fat under their hide than anybody else. For male Inuit, this would be counterproductive: Body fat is correlated with a lower thyroid setting, which in turn lowers the metabolism, which in turn lowers body heat. Female metabolism seems unaffected by body fat, probably because females need to carry so much fat in order to reproduce. Besides, it's muscle tissue, not stored fat, that converts calories into the metabolic heat to warm the cockles of your heart. Which is another reason that females chill easier than males—their muscles are smaller.

I suppose I shouldn't be surprised at the variety of responses humans have evolved to the challenge of cold and winter. First of all, we've spread to dozens of cold ecosystems, each of which applies its own set of pressures to an animal. Second, nature abounds in cold-combating techniques. Take the ground squirrel of Canada and Siberia, who adjusts his metabolism to one extreme: He dials back the thermostat to barely above freezing, curls up in a grass-lined den, and hibernates for seven months of the year. Poised at the other end of the spectrum, and eating constantly, we find the least weasel, whose habitat circles the North Pole. Rather than economizing, he burns calories at a rate befitting an animal five times his size. He must be on the verge of spontaneous combustion. And the polar bear takes cold survival in yet another direction: He's so big and insulated with fat that overheating regularly drives him to cool his belly on the ice.

Interestingly, even without a cold crisis, northern humans in comfortable shelters tend to burn hot, a bit like the least weasel. Their basal metabolism is quite a bit higher than mine (reports range from 3 percent to 30 percent higher). So Ilya may be eating his way to warmth even on a normally cold day like today. He doesn't look like he's eating a lot. Inside his blue tent he offers me a battered enamel bowl of soup from a pot on the stove. A lean ptarmigan leg swims in broth with a few tundra herbs. Ilya's staple food, caribou, is also lean. Without the chubby seals and fish that the Inuit enjoy, the Nenets probably can't afford to let the metabolic fires rage. I feel silly eating Ilya's lunch when I have a kitchen at home stocked with enough calories to maintain his family for a week. But my guide assures me we are paying for Ilya's hospitality, and so I wrap my clammy fingers around the red and white dish, and inhale. It smells of clean tundra, and it warms me from the inside, too.

Ilya's caribou suit hides his shape. But if Carl Bergmann was right, underneath that caribou fur is a block of a body. Bergmann, a German biologist of the 1800s, noticed that animals are generally larger the farther north you go. It's a sensible theory. For one thing, an abundance of muscle cells produces an abundance of heat. Furthermore, large animals enjoy a high ratio of warm innards to cool surface, which means they hold heat better than small animals. The perfect shape for an Arctic animal would be a sphere—the shape that minimizes surface area, and the shape most animals take when we curl up against the cold. Thus in accordance with Bergmann, a male white-tailed deer, whose range runs from southern Canada to Colombia in South America, weighs just 110 pounds (50 kilograms) in Venezuela, but tops out at 440 pounds (200 kilograms) at the northernmost limit of its range. The human animal conforms, as well. On the face of it, I'm much larger than Ilya. But I had the benefit of prenatal care, vitamins, antibiotics, and food free of parasites, and this sort of disparity messes up the comparison of a single *Homo sapiens* to another. Nonetheless, when scientists measured one hundred populations of humans from a variety of locations, they found that Bergmann rules: On average, north-adapted humans are bigger than tropical races. Bergmann's power is not absolute. In some animals, the effect is tiny, and some species refuse to comply altogether. As Bergmann himself complained, "It is not as clear as we would like." But that's nature for you. There is always more than one ecological pressure bearing down on an animal, nudging its body hither or yon.

Also under Ilya's tunic, I'd guess that Allen's rule is in effect: Ilya's arms and legs should be shorter than mine, relative to our torsos. Thirty years after Bergmann issued his rule, American naturalist Joel Allen chipped in a related decree. Northern animals shall display shorter limbs, quoth he. Allen's reasoning was the same as Bergmann's—when you reduce surface area, the animal is better able to maintain heat. Limbs are long expanses of surface, with scanty innards to heat them. The contrast between the northern snowshoe hare and the rangy rabbits you'll find in deserts is classic Allen: The snowshoe hare has shorter legs, shorter ears, and a more spherical head. The Arctic fox compares similarly with his red cousin of temperate zones: In addition to the world's warmest fur, the Arctic fox has stumpy ears and legs.

How Ilya's face serves his life in the cold is more questionable. Once upon a time, anthropologists swore allegiance to a "cold-engineered" rule for Mongoloid features. Under this rule, the barely there browridge (the bone the eyebrows sit upon), the epicanthic fold of fat in the eyelid, the arching cheekbones, and the pad of fat on the cheek all reflect genetic adaptation to cold. And it sounds sensible. The net effect approaches the spherical ideal that minimizes surface area.

However, investigators now suspect the Mongoloid face is largely a case of "genetic drift." There's more than one way to evolve a distinct feature. Natural selection is the obvious way: A protruding nose begets frostbite, gangrene, and death, hence the gene for a protruding nose dies out. The small-nose gene remains. But another way to develop a distinct feature is to strike out for new territory with a small number of humans. Over the generations, the little scoop of gene soup you all represent will be boiled down and concentrated into a particular look shared by all. Later, if you should merge with another group, your features will start drifting off in a new direction. With research stubbornly refusing to demonstrate that the Mongoloid face fends off frostbite, genetic drift emerges as the most likely explanation. Recent DNA analysis is bolstering the case. The entire human genome—the genes that build a human—evolves at a predictable rate as beneficial mutations strike and are incorporated. Thus any feature that evolves faster than this background mutation rate is probably propelled by an environmental pressure. But measurement of lots of human skulls shows that the Mongoloid skull has changed shape no more rapidly than the entire genome. This implicates drift, not natural selection. So it's possible that all humans with the Mongoloid skull shape share a small and ancient group of ancestors whose heads, for no particular reason, acquired that shape.

The debate isn't over. The very latest word on the cold-engineered face is that perhaps two—just two—of its features are indeed naturally selected. By measuring 2,472 skulls from Africa, the Middle East, Europe, Siberia, eastern Asia, and South America, an anthropologist from Stanford University concluded that, of all the faces in the world, only the cold-adapted face shows any evidence of evolutionary engineering. He found that the Buryat of Siberia, a Mongoloid population, do in fact have a wider (more spherical) head than genetic drift would account for. Furthermore, their nose—tall and narrow—falls outside the drifting range. These features, then, are probably the result of natural selection. The advantage of a spherical head is evident, but what's the point of a tall, narrow nose? I'm stunned to read that a human body that's inhaling –25°F (–32°C) air spends a whopping one-quarter of its total energy budget just to preheat that air so it doesn't shock the lungs. And all that warming happens in the nose. A narrow nose, nestled in the face, will heat air more efficiently than a wide one that pokes out in the cold. But before we toast the Mongoloid skull and nose as a case of evolution in action, we should note that the author analyzed only one group of northern humans, and is himself quick to warn that it's only a start.

As a Russian helicopter whacks into view to extract me from the Stone Age, the guide slips Ilya some Russian banknotes and a bag of hard candy. I wince. I don't believe there's a dentist within five hundred miles. But who am I to dictate the proper balance of joy and pain in another human's life? Ilya waves as the chopper rises, his teeth brown, his eyes sparkling black beneath their epicanthic folds.

COLD AND DAMP AND DISMAL HABITATS

Heading south of the Arctic Circle on my journey, I stumble almost immediately over Iceland. It's northern, it has snow and glaciers, but it's different. Whereas the Arctic is cold and dry, certain unfortunate spots on Earth are cold and wet. There is no environment I prize less. Even in summer, Iceland can break your heart. From a volcano's shoulder you may stand and gaze down at a plain of pasture so green it seems to vibrate in the sun. Beyond, a blue sea glitters. You peel back a layer of fleece and take the sun on your chest. And then rain comes out of nowhere, attacking in horizontal tirades over the stony earth to sting your face. Perhaps it will turn to snow. Or sleet. Either way, it will continue to strike from angles that no reasonable human would anticipate, driven by a manic wind.

I have not visited Iceland in the winter, and the truth is humans haven't lived there long enough to evolve a specific response to sideways rain. Vikings migrated to Iceland only about fifty generations ago. To evolve a distinct trait so quickly would be impressive, but not impossible. Recently crossing my desk was a report on cane toads introduced to Australia seventy years ago. In seventy toad generations, some evolved longer legs, which carried them to new territories faster than the original, short-legged toads.

The Vikings are to my liking: Even the prettiest of the blondies seem equally capable of riding a horse or changing a flat tire on a mountain road in a blizzard. The citizenry all appear to stay up for the entire summer, then spend the winter reading and writing books. Again I was on a travel assignment and again the local culture presented its particular hazards—my photographer, lying out in his sleeping bag one sunny 2 a.m., came very close to getting his nose pinched by a feral pack of gorgeous teens. But, charming as the Vikings are, they do wear wool and GORE-TEX, and their shelters are warmed with heat pumped from the volcanic rocks beneath their feet. To meet the true masters of adaptation for cold wetness, we must journey to a similar latitude, in the opposite hemisphere. Here dwell—or dwelled—a race of humans for whom my esteem is boundless. Inhabiting the southern tip of South America, where another battering wind carries rain, sleet, and snow, they were the Tierra del Fuegians.

Darwin's description upon meeting these humans when sailing through is unforgettable. To set the scene, this is not a restful place. Winter temperatures stick close to the freezing point. A hot summer day might reach 64°F (18°C). It's absurdly windy and apt to produce squalls of rain at any moment. Darwin had already expressed wonder over the area's eastern tribes, who braved the climate attired only in a guanaco cape. (The guanaco is a small, wild version of the llama.) He was further stirred by their central-region neighbors, who faced the elements wearing a scrap of otter fur "the size of a pocket handkerchief." But when Darwin met the canoe humans of the southernmost (cold-most) islands, he was undone. The humans here spent their days diving for seafood, and fishing and seal hunting from canoes. Year-round and naked.

Wrote Darwin, "A woman, who was suckling a recently-born child, came one day alongside the vessel, and remained there out of mere curiosity, whilst the sleet fell and thawed on her naked bosom, and on the skin of her naked baby!" Poor Darwin didn't yet realize that the Fuegians could have been Exhibit A in his forthcoming theory on natural selection. They had been adapting to this habitat for ten thousand years (four hundred or five hundred generations). "At night," he wailed, "five or six human beings, naked and scarcely protected from the wind and rain of this tempestuous climate, sleep on the wet ground coiled up like animals." How could he have known they were cozily hypermetabolizing?

Like Darwin, I'm fascinated with the Fuegians, who are now culturally extinct. I burrow into the flaking memoirs of E. Lucas Bridges, the son of missionaries, who grew up in Tierra del Fuego in the late 1880s. Bridges's playmates, members of an inland tribe called the Ona, wore the guanaco hide—barely. A male used his left hand, which also held his bow and arrows, to grip the skin closed at his chest. This was so he could quickly ditch the skin when an opportunity arose to dance, wrestle, stalk game through the snow, and so forth. Bridges, who often traveled cross-country with a few Ona, never lost his wonder at their ability to slog through snow, rain, and rivers clad in just a robe and moccasins, then at night wring out the moccasins, put them back on, lie down in the guanaco hide, and snore soundly. He did say that when the Ona broke a trail through deep snow, the lead man would don leggings made from guanaco leg skin to protect his shins from the icy crust.

The females were equally hardy. The tribes were nomadic, as hunter-gatherers tend to be, and the transport of tools and offspring seems to have fallen to the females. Not that there was much to carry besides children. The Ona took a central Australian approach to shelter, leaning a few branches together to break the wind.

The Fuegian physique plays a counterpoint to that of Ilya. Some South American tribes, including the Ona, did in fact tower over the first European migrants from warmer latitudes. This could reflect Bergmann's rule, but it may also have resulted from a stunting effect in the Europeans, whose home range was crowded and foul. It's the Tierra del Fuegians' head shape that's puzzling. Flouting both Bergmann and Allen, their skulls are not spherical, as a cold environment would predict, but elongated like my own European model. How environment and human head shape relate is obviously a puzzle with a few pieces still missing. The heads of female Fuegians further bedevil the question: They had the most heavily built skulls of any population ever measured. Bridges's description of domestic violence among the Ona may have bearing: A male who judged his mate to be out of line was justified in shooting an arrow into her leg or clubbing her. Any time a male became wrathful in camp, all the females reflexively dove to the ground and pulled guanaco robes over their heads. A sturdy skull may have paid handsomely for a female Fuegian.

It's with regret that I leave these humans, both the Fuegians and the Icelanders. Of all the humans who have penetrated all the corners of our range, these—the cold and wet humans—seem toughest to me. A squirrel who spends seven cold months in a voluntary coma is remarkable. But a human who faces far worse weather and *doesn't* take refuge in a coma is more so.

HABITATS AT DIZZYING ALTITUDES

To continue my ramble through the human range, I'll turn closer to home. Visible on a clear day from my town is a hulking mound on the horizon. Mount Washington, president of New Hampshire's White Mountains, stands 6,288 feet above sea level. This is about 6,258 feet above my personal home range. Some humans from the American West categorize this mountain as piffling. My sister, who haunts the Sierra Nevada, wonders if Washington is even a mountain at all. But it's the best we have around here. When we climb the surrounding peaks, we orient ourselves by its noble profile. And on a fine summer day, with a few comrades, I climb it.

Approaching the summit, walking across a stony field of dwarfed flowers, I feel fatigue I'm unaccustomed to. I'm not muscle tired. But my heart is knocking a bit. I feel light in the lungs, with an urge to yawn. And at the peak, where the wind rushes among the rocks, I hop onto a boulder and my heart knocks harder still. The cause is the air one mile up in the sky. It's only about 80 percent as dense as at home by the sea. So each breath nets only 80 percent of my customary oxygen. My lungs have to fill more frequently or my heart needs to work harder if they hope to deliver the usual amount of oxygenated blood to my tissues. They're managing. I have no headache, no nausea, no malaise. But bear in mind I'm standing on an anthill, compared to some of the places where *Homo sapiens* has adapted to live.

If I were to hike my lungs up to Lhasa, Tibet, I could probably work up a world-class headache. And when I slept, my respiratory rate would fall, suffocating me back to wakefulness. A new bit of research suggests the "good bacteria" in my gut might slowly be overrun by toxic varieties. I might hike right into serious altitude sickness. Seasoned climbers who ascend too quickly take a little headache and vomiting in stride. But this can progress to edema of the lungs—climbers are counseled that the sound of a crackling paper bag is most likely caused by your lungs filling with fluid, and it's time to go down. Coughing up pink foam indicates that you should have left a little earlier. Brain edema is less forgiving, and announces itself with the loss of bowel function, then coordination, then consciousness. Either edema is an effective method of flicking a body out of the gene pool.

But if I didn't go too high too fast, my flexible physiology could adjust to life with less oxygen. The headache and nausea would fade. Within a few days, my breathing would grow deeper, sucking in more oxygen per inhalation. My blood pressure would rise as my arteries constricted to push blood into the farthest reaches of the lungs to scavenge oxygen. My bone marrow would accelerate the manufacture of red blood cells, those freighters of oxygen. And the right ventricle of my heart would bulk up from pumping all the fresh blood through my constricting arteries. Eventually, I would be able to get up and be productive.

I could have acclimated even better, had my parents sent me to Lhasa as a kid. I would have grown a barrel chest to accommodate bigger breaths. I can't find any research on whether that adaptation is a matter of special gene expression—whereby the environment would flip the switch on my big-chest gene—or just normal bone remodeling. My bones are forever changing their shape, based on vibrations they absorb from my activities. If I took up tennis, for example, the arm bones on my racquet side would actually grow thicker and longer. But it's too late for me now. A supersized chest has never been found in studies of lowlanders like me who immigrate skyward in adulthood.

So, beyond beefing up the heart, blood, and respiratory rate, there's not much else a visitor to high altitudes can do. As a female, I'll probably suffer less and adapt faster than a male. Although the reason isn't known, research on rats suggests that female sex hormones work some protective magic. It also makes clear that male rats die like crazy at high altitude, while females scamper away unscathed. Despite the popular belief that human females have a similar advantage, scientific evidence is lacking.

Had I truly burned to conquer the world two miles up, ideally I would have had ancestors who came to Lhasa thousands of years ago, bearing stone tools and chasing yaks across the wide, bright steppe. (Whether humans came five thousand or thirty thousand years ago is debated.) I'd be sitting pretty now. The Himalayas intrude high into the atmosphere. Lhasa, in a narrow river valley between mountain ranges, lies at 2.3 miles (3.7 kilometers), or about double the elevation of my Mount Washington. Oxygen is just 60 percent to 65 percent as plentiful as at sea level. (It's rarer still at the world's highest town, Wenzhuan, almost another mile above Lhasa.)

Mountains are the only sector of the human range where Stone Age tools would have been worthless against the environmental challenge. Whereas early humans who migrated into the Arctic could add another layer of fur and light a fire in their snow caves, there is no cure for thin air. Even today's best high-altitude tools—metal tanks of oxygen and a mask—are clumsy and primitive. The only way for early humans to expand their range upward was to evolve. Tibetans have done this.

It's a fruitful place to study the effects of altitude, because living cheek-by-jowl with the indigenous Tibetans are throngs of Han Chinese immigrants. This race came uphill only after the Chinese military invaded in 1951. Politics aside, human physiology is pretty eloquent on the question of which bodies "belong" in Tibet. The Han will require thousands more years of good mutational fortune to catch up with the Tibetans' current suitability for life in the sky.

Compared to their Chinese neighbors, Tibetans have bigger chests, with bigger lungs in them. Even a Han child who grows a rounder chest during a childhood at this elevation will never achieve the lung volume of a native Tibetan. Tibetans also have naturally more powerful hearts—only 17 percent of Tibetans display an enlarged right ventricle of the heart (a sign of overwork), compared to 29 percent of Han immigrants. And even compared to other races of mountain-dwelling humans, Tibetans manage to keep their brains fully oxygenated, even when they're exercising and all their muscles are calling for air.

The genes get more impressive as you go higher: One research team recruited twenty super-high-elevation Tibetans and brought them a half mile (.8 kilometers) down to Lhasa to compare their physical performance with the Lhasans'. The high-elevation Tibetans, pedaling stationary bicycles, were able to get 17 percent more muscle work out of the same amount of inhaled oxygen. And even as they whizzed off to nowhere on the bikes, they were breathing more slowly, and their hearts weren't pounding like those of the "low-altitude" Lhasans.

How do these superhumans do it? For one thing, Tibetans breathe more often than lowlanders. This refreshes their lungs with oxygen more frequently. But they don't have fleets of extra blood cells to distribute that extra oxygen. In fact, some studies show their blood is even less oxygen-saturated than my sea-level blood. For now, how Tibetans feed their brains and muscles on thin air is a delicious mystery. But in my hunt for clues, I stumble across a particularly piquant piece of research.

Cynthia Beall, an American anthropologist, surveyed blood-oxygen saturation in Tibetan females, then grouped the females by family to see if a genetic component might emerge. Her data imply that some Tibetan females express a gene that saturates their blood with a 10 percent bonus of oxygen. And here's why it matters: These females are five times—five times—more likely to bear offspring who will survive infancy. To put it more grimly, the infant mortality rate for females who *don't* express this (presumed) gene is five times higher. Both types of females produce the same number of babies. But the offspring of the oxygen-saturated females are far, far more likely to survive and reproduce. This is the beating heart of natural selection at work. And with such a grisly infant-mortality rate, that low-oxygen gene should be fading fast.

And speaking of offspring, here's a tip: Don't gestate one at high elevation if you don't have the genes for it. Comparisons of Tibetan and Han Chinese infants born in Lhasa show that Chinese offspring are notably smaller, and despite an elevated blood-cell count, they're unable to keep their bodies fully oxygenated. They're prone to turning blue.

Tibetan females have evolved a different response to elevation than males have. Cynthia Beall has analyzed the oxygen saturation of Tibetan males, as well as females, and discovered that when males reach their twenties, their blood-oxygen levels start to falter. But Tibetan females seem immune to that aspect of aging. Their blood stays fully aerated through their forties. Once again, I would boldly speculate that a breeding-age female has special needs: If she can't keep her fetuses aerated, they won't repay her by living to carry her DNA forward.

Now, one might think that if you've seen one population of high-altitude humans, you've seen them all. But as with cold adaptation, humans have found more than one way to skin a cat. Humans in the Peruvian Andes have taken a different evolutionary path. Unlike Tibetans, Andeans do have tons of additional blood cells, and that blood can carry an unusually heavy load of oxygen. Andeans do not, however, appear to have mastered the art of gestation as Tibetans have. In the mountains of Peru, the higher you go, the smaller the babies are. And no matter where you find them, small babies are fragile babies.

And lest we forget the Ambaras humans of the Semien Mountains in northern Ethiopia—okay, I never heard of them either—Beall has not. Neglected by scientists until Beall arrived, these humans live more than 2 miles (3.5 kilometers) in the air. How long they've been evolving there is unknown, but it's been long enough for them to make some modifications to the human plan. And these adaptations represent yet a third pattern. Like Tibetans, these humans have no more than the normal number of blood cells. But somehow, they're able to load up these cells with extra oxygen. Their saturation is much higher than either the Andeans' or the Tibetans'. The changeability of DNA thrills me. A single challenge—elevation—has produced three separate evolutionary adaptations.

Other high-altitude animals are no less thrilling, I suppose, in their evolutions. A Japanese team has tested the high-altitude pika, a winsome microbunny, against the lab rat, to determine how the pika handles thin air. Researchers put the two species in a decompression chamber, where they could re-create the environment of any altitude. The rats, when the chamber simulated a climb from 2,260 meters to 5,000 meters, developed high blood pressure, as the thick walls of their arteries constricted. The right ventricles of the rats' hearts swelled from overwork, and they produced more red blood cells as the oxygen dropped. The pikas retained light and flexible arteries, which left their blood pressure at ease. Another alpine species, the Andean llama, has evolved an alternate response. Llamas have small blood cells with special chemistry that allows them to soak up oxygen quickly. And yaks, those shaggy, mountain cows of Asia, apparently maintain their tranquil blood pressure by remodeling the very cells that form the walls of their arteries. (And in a trick reminiscent of the Australian Aborigines, yaks can also dial back their metabolism to save energy when the mountain air grows chilly.)

My second visit to Mount Washington took place in winter and lasted for three days. Yes, once again I was on assignment, this time at the mountaintop weather observatory. From this bunker a skeleton crew of humans makes hourly reports of wind (record setting), snow (carried by record-setting winds), cloud cover, temperature (quite low), and so forth. Again the local culture was a wonderment—geeks this time, severe and devoted weather geeks who would ride the snowcat up to the summit but bring along sleds for the descent when their weeklong shifts were done. My three days were sufficient to start my flexible anatomy acclimating. The right ventricle of my heart, pumping iron twenty-four hours a day, bulked up. My blood pressure crept up, too. And my bones accelerated their production of blood cells. It happened too gradually for me to notice, until the day a storm threatened and I had to either hike down in a hurry or commit myself to another few days in the bunker with the geeks. Thanks to the human genome, which seems prepared for almost any strange place I might want to swim, crawl, migrate, or climb to, I was prepared to take flight. Scuttling down the mountain among the first whipping flakes, I was not as oxygenated as a Tibetan or a yak, but perhaps I was a bit more sprightly than a male rat.

THE HIDEOUSLY HOT HABITATS

From the mountains of Tibet, let's tumble downhill, to a desert that lies in the rain shadow of the Himalayas. The great wall of those mountains pokes high into the atmosphere to rob the clouds blind. Moist air rolling north from the Indian Ocean has to climb so steeply to cross the Himalayas that it cools and drops all its water. Once over the wall, the wrung-out air has nothing left to give. The Gobi, a yellow-tone desert that consumes swaths of Mongolia and China, is one result.

I was deployed to this vast, beige corrugation to write about the annual dinosaur expedition of the American Museum of Natural History. For days that blended into nights and new days, our dusty convoy struggled across the roadless desert toward the dig site. We were eighteen humans of diverse race and interest—some dinosaur hunters, some mechanics, some geologists—and seven vehicles of comparable diversity—two German and two Russian SUVs, two food-and-fossil trucks, and a fuel truck that could get bogged down in the amount of sand I shook out of my underwear each night.

The desert presented one of the easiest habitat adjustments I ever made. When I go north, my fingers go white and numb. When I visit the humid tropics, my body swells and my hands throb for two days. But in the daytime, at least, the desert is closer to the temperature *Homo sapiens* is built for. It's not quite perfect, because afternoon temperatures frequently climb thirty degrees above the ideal of about 80°F (27°C). But with sufficient water, I adapted with only moderate misery.

The direst threat to my life in that particular desert was boredom. Because human tools still founder in the face of dust and heat, the entire convoy coughed to a stop every hour or so, when one of the tools overheated or sank in the sand. With nothing to read, nothing to look at, and nothing happening, I spent hour after hour sitting with my back against a tire, glaring back at a hot sky. Sometimes, as I scanned some new valley from the discomfort of my tire, a white dot would blink back at the base of the next row of charcoal mountains. Humans.

My eyes could make out nothing except the spot that must have been a circular *ger*, or yurt. But in the vicinity of that shelter I knew there would be a herd of sheep, a handful of camels, a dog or three, some of the most unkillable horses on the planet, as well as some of the most unkillable humans. And there must be water, perhaps in the form of last winter's snow percolating down through the fractured mountain.

Desert ecosystems are defined by their extremes: They're extremely hot in daytime, shockingly cold at night, and most notably, they're dry, dry, dry. Only a handful of plants and animals have evolved the means to wring a living from desert habitats. For humans, heat and cold aren't the deal breakers. It's the aridity, and the resulting biological poverty, that test our mettle. Our warm-blooded metabolisms demand the regular shoveling-on of fuel, which in deserts is sparse. Moreover, we're uncommonly moist animals, with water always oozing from our hides. Although some humans have evolved to stretch the distance between watering holes, even the best of us is not a genuine desert animal.

But let's begin with the easy stuff—not frying. As a quasi-northerner, I'm not a grand example of this, but I suppose I provide a cautionary tale of maladaptation. For starters, I'm not stick thin. Recall that spherical is the ideal shape for a northern animal, so as to minimize surface area. The opposite is true if you hope to shed heat efficiently. Like the radiators that give heat to my house, you want lots of fins and skinny tubes for heat to ripple from. And hence the classic hot-region body of the Maasai, the cattle-herding humans of Africa's southern drylands. The Maasai are as elongated as Giacometti sculptures, with narrow trunks and legs that go on forever. Body fat is one of the greatest hindrances to staying cool, and the Maasai look as though they've done away with it completely.

I, by contrast, am no Giacometti sculpture. But for that matter, neither are the desert Mongolians. Their shape has perhaps struck an evolutionary balance between the furnace of summer and the deep freeze of winter, since their desert is located farther from the tropics than most. Mongolians are proportioned more like me than like the Maasai. I wouldn't be surprised, given the quantity of sheep fat in the Mongolian diet, if they're even able to hypermetabolize when they're cold. The preternatural strength of the expedition's Mongolian mechanic, Timur, suggested that there may be an excess of hot-burning muscle under that skin. Six inches shorter than I am, and a few inches narrower, Timur would amuse himself by lifting me suddenly off the ground, convulsing with laughter as he did so. In terms of science, I can find none relating to the Mongolians' physiology. Biological scientists are generally more interested in Mongolia's endangered desert dwellers—the Bactrian camel, Przewalski's horse, the snow leopard—than its humans.

Regardless, even we humans who aren't elongated have ways to stay cool when we migrate out of our comfort range. Our furlessness, as we saw in chapter 1, is probably key. Instead of growing a woolly coat to insulate us from incoming heat, the entire human race opted to scrap the fur and rely on water cooling. Even when human skin is cool, it leaks moisture. Our leaking rate is similar to that of a tropical caiman (alligator), who never strays far from water.

Such a soggy skin can be a serious hazard, but as long as we're able to replenish the moisture and salt we ooze, our cooling ability is spectacular. Humans can sweat more than any other animal—nearly a gallon an hour, if need be. In fact, the more days I spend in the Gobi Desert, the more profligate I become with the sweat. My body's temporary solution to excess heat is to throw more water at it. The long-term solution, however, is out of my reach. Bushmen adapted to Africa's Kalahari Desert, and Australia's desert Aborigines (and probably other desert dwellers who have evaded scientists' stationary bicycles) have evolved low sweating rates, and they stay cool just the same. This water-saving trick is available only to those whose ancestors have spent many generations evolving in a habitat so dry that it kills the leakiest and rewards those who perspire the least. Tropical humans also retain a functioning gene that regulates the amount of salt they lose in sweat. Populations farther from the equator, however, like my own ancestral humans, let this gene go to seed, and our ability to retain salt is patchy. That's a problem.

Even the best among us will succumb promptly if we can't replace the water and salt that inevitably escapes our hide. A hot body will continue to gush water until it either cools off or collapses. Even a human who has unlimited access to water can sweat himself to death if he can't replace his lost salt. In a charmingly simple experiment, a scientist in 1933 took a long walk in the desert with a dog. Both drank as much as they wanted. At the end of the day, the dog had lost only 0.5 percent of his body weight; the human was down 3 percent. The human's sweat poured sodium out onto his skin. As his body lost salt, it lost the capacity to retain water. But the dog's panting spent only water, and the electrolytes he conserved kept his water inside, too.

As I sweat in the Mongolian sun, my body quietly adjusts in other ways. Although temperate-region specimens like me are burdened with thick arms and legs, at least our circulatory systems can open wide, the inverse of their cold-conditions clampdown. By flushing blood to my extremities, I can dump heat into the air—as long as the air is no hotter than my blood is. As the days pass, my body starts sweating earlier in the day, as a preventive measure against overheating. I expend less salt in my sweat and urine. My blood, spread thin to perfuse my skin, bulks up on water. This calms my heart rate, which stabilizes within a few days. My temperature stabilizes, too. After two weeks, I'm so accomplished that my body no longer resorts to surface blood cooling, and my blood volume returns to normal. The dust does its best to plaster over the sweat pores in my skin. But when my erudite companions blab about mandibles and maxillas around the campfire each night, I find I have time to rub little balls of grime off my hide. I've acclimated.

Naturally, any animal whose species has spent eons evolving in the heat is better off than I am. Occasionally our convoy crosses paths with a handful of camels. Bright tufts of yarn in their mane identify them as livestock, not wildlife. Narrow enough to be shaded by their own fatty humps, the camels cast weird shadows. Long legs raise them above the hot sand. Their fur is surprisingly thick and prevents the 110°F (43°C) air from penetrating the 100°F (38°C) camel. Camels also allow their body temperatures to fluctuate more than mine can without inviting catastrophe. In the heat of the day, a camel runs what for me would be a fever of 102°F (39°C); in the chill before dawn, it flirts with mild hypothermia.

Goats and sheep, staple foods for desert humans, manage their heat differently, through brain cooling. Like a number of animals—deer, camels, and rabbits among them—goats and sheep can adjust their circulation and pant through the nose and mouth to selectively cool their brains below body temperature. Initially, scientists presumed this was to prevent a meltdown of that most vital organ. But increasingly, it looks as though the brain just makes a marvelous radiator for the rest of the body. Experiments have shown that without brain cooling, goats would need half again as much water to keep up with a soaring sweat rate. By shedding heat from the brain instead, they lose less water. Whether I possess the ability to selectively cool my brain is the subject of a long-running scientific debate, with no clear evidence either way—which probably means that if I've got it, it's nothing to write home about.

Some animals have evolved to stay cool by staying small. In keeping with Bergmann's rule, desert fox species are generally smaller than their temperate-realm cousins. In my northern neck of the woods, a red fox weighs an average of 11 pounds (5 kilograms). But the kit fox from the deserts of the American West weighs about half that. And the Saharan fennec fox tips the scales at just 2 or 3 pounds (1 to 1.5 kilograms). In addition to shedding heat easily, such dainty bodies also require less food and water.

And, of course, most desert animals are maestros of moisture. Kangaroo rats of the American Southwest can produce all the water they need as a by-product of metabolizing the seeds in their diet. Australia's zebra finch is equally frugal. I've read an account of a scientist who tried to determine how long this dapper creature can go without water. He threw in the towel after a year, saying he felt sorry for the bird.

I, too, sometimes suffer heartache for the evolutionary fate an animal endures. I'll never forget the photo of the "dew bug" I saw in *Nature* in 2001. The *Stenocara* beetle, native to Africa's driest dunes on the coast of Namibia, can't hope to commute to a watering hole the way a bigger beast might. And so on those rare dawns when a fog wisps in from the sea, he tilts his nose down and unfurls his bumpy wings. Moisture condenses on the bumps, and this niggardly film slides down troughs in his integument to reach his mouth. It kills me. I would like to hunt down every one of these beetles and give them hamster bottles. I understand that as we animals evolve our quirky adaptations, we also evolve a capacity to endure our fates, but still.

IT'S NOT THE HOT HABITATS, IT'S THE HUMID ONES

When my deployment to the desert was done, I returned to a city, washed my dusty "tan" down the drain, and began to de-acclimate. After a month at home, my hard-won gains were erased. And soon it was time for some new trial by fire. Or, more aptly, by fungus.

From Mongolia, let's drop south past the equator, and then travel a quarter globe west, to Madagascar. On the western side of this big island lies a national park composed of rain forest, rain, spiders, forest, rain, fungi, and soaking, sodden forest. Oh, and heat. Plenty of heat. This time, I'm writing about a herpetologist who's surveying the entire island for reptiles and amphibians. Because these critters estivate in the dry season, we have to visit when they're out and about, which is the aptly named rainy season. So our muddy party—

the British herpetologist, four Malagasy researchers, and half a dozen porters—is camping for ten days in the rainy rain forest. By day we canvass the steaming jungle for geckos, snakes, and iguanas. By night we do the same. Periodically we eat rice boiled with curry powder and canned sardines.

The greatest difference between desert and jungle, for a visiting human, is the humidity. If deserts are defined by their desiccation, the forests of the tropics represent the opposite extreme. My dry, northern corpus registers the contrast painfully.

In the first days, when my blood rushes to the surface to cool, my body dumps extra water into the blood, and my flesh swells like a sponge. This heat edema makes my hands so puffy they throb. "Go home," my laboring heart drones, pushing all that water. "Go home, go home." But the same adaptations that serve in the dry heat of Mongolia eventually muster themselves in Madagascar. The hands deflate. The heart rate stabilizes. The sweat starts earlier—and pours forth even more profusely than it did in the desert. The humid air is too saturated to absorb my sweat. My skin remains gummy with perspiration for the duration of my stay. Here in the tropics, the heat holds sway day and night. Like Darwin's Tierra del Fuegians, I could lie down on the ground at night and sleep comfortably without shelter if it weren't for spiders big enough to bundle me up and suck me hollow.

Human Pygmies may represent the ideal human adaptation to hot humidity, according to many anthropologists. The distribution of these small races is decidedly moist. They dwell in the forests of central Africa, the Philippines, the Malay Peninsula, and the Andaman Islands. Their small size presumably eases the chore of cooling and makes them fuel efficient in jungle ecosystems, which can be surprisingly short on chow. Like Tierra del Fuegians, many Pygmy humans could measure their entire wardrobe in square inches, spiders be damned.

Once again my physiology exposes my temperate genes. I'm big. Just walking in tropical forests is complicated by my long arms and legs, my towering head. And I'm relatively bulky. Like a polar bear, I cool off slowly. Here in the cradle of my primate ancestors, I suffer. My skin, translucent and sticky, chafes against itself—I am, I may have mentioned, too bulky for this. Fungi drill into my toes, disabling me. I am always, always thirsty.

My species can exploit a range broader, and higher, than almost any other creature, thanks mainly to an affinity for tools. Without fire and clothing, we would be confined to a tropical girdle around the planet's middle. But tools, plus an ability to extract food and shelter from a ridiculous assortment of ecosystems, open up the latitudes as well as the altitudes. The human animal can live almost anywhere.

And yet, despite how easy migration has become for humans, most of us stay in the ecosystem where our ancestors evolved. So I head home. Home for me is high enough on the globe that the length of the days surges, then wanes, with the seasons. As winter expands into spring, then summer, my environment changes slowly enough for my body to keep pace. Never do my hands ache. Just gently, one June day, I'll notice that my wedding ring, loose all winter, is now snug. And then, although the first day of autumn cold may nip, the second one will merely tickle, and before long I'll be shoveling snow in a light sweater, heedless of the weather.

That's home, halfway between the North Pole and the equator, on the east coast of North America. There each cubic foot of air weighs 1.2 ounces. About forty-two inches of rain and snow fall each year. Summer noons are 25°F (–14°C) cooler than in the Mongolian Gobi. That's roughly the latitude where my ancestors evolved in Europe. And that is where I function best. Thanks to the prehistoric restlessness of my species, I have inherited a body flexible enough to survive the snapping-cold Arctic and the toe-rotting tropics. I'm impressed. I'm grateful. And now I'm going home.

5
A DOG IN THE MANGER:
TERRITORIALITY

The human is territorial for the same reasons many other animals are: Familiarity with the ecosystem speeds the daily task of securing food and shelter, and territoriality lowers the risk of violent conflict. By marking boundaries with fences, plants, stones, plastic birds, and other objects, a human signals to neighboring primates that he won't attack unless the line is crossed.

Due to the social nature of this animal, a territory need not provide for all the resident human's needs. In fact, a group of humans will often cluster their individual territories around shared resources such as fresh water, hunting grounds, or a highway. (This is contrary to most territorial animals, whose territory must provide for all their needs.) Still, the quality of a territory has far-reaching effects, because a poor territory—one that's noisy, crowded, and lacking in plant life—will undermine the health of the resident human.

When the human territory is threatened with invasion, the resident animal enjoys the same home-field advantage demonstrated by sparrows and other territorial species. Thus, even a fairly weak human often can prevail on its own turf. Humans also exhibit the same hierarchy of defensive behaviors shared by many animals when the territory is invaded by a more powerful competitor. In spite of the home-field advantage, the first choice is usually to flee. But when a human is cornered, he'll opt to hide, then threaten, then attack, as the invader closes in. The female, usually smaller and hence more vulnerable to violence, is typically quicker to abandon her territory than is the male.

Human territoriality has been altered by the phenomenon of densely clustered shelters. Back when a human territory met all the needs of its inhabiting clan, conflict was limited to warfare at the borders or during invasive campaigns. Now most humans claim minuscule territories and travel into shared areas to forage. As a result, the territorial conflicts of this species may be growing more frequent, but also more benign, since they tend to flare over parking spaces and Tickle Me Elmo dolls rather than control of a life-giving water hole or of breeding-age females.

THE SIZE AND SHAPE OF A HUMAN TERRITORY

The crow family that owns my yard oversees a total territory of about eight city blocks. They wouldn't mind having more acreage, but that would obligate them to defend longer borders. So, because these eight blocks provide sufficient food, tall lookout trees, and dense evergreens for sleeping and nesting, they limit their empire. On occasion, my crows (a mated pair and some grown children) sneak into the neighboring territory to see if the worms are meatier there. If discovered, my crows are assaulted and driven home in a hullabaloo. So generally, they stay home and save their energy for baking the next batch of crows. And that, in an eggshell, is the reason so many animals are territorial. When everyone agrees on the boundaries, and believes there will be trouble if they cross them, then everyone also spends less time fighting.

For a quick vision of how life might be if humans were not as territorial as the crows, imagine that you do not have any formal right to your customary abode. Nor does anyone else on your block. When you all leave work at the end of the day, there will be a mad rush for the various shelters. Naturally, the majority of humans will want the most comfortable and safe ones. At the door of a high-rise with a doorman, a horde of primates will battle for first entrance. Cars will scream out into the suburbs to collide in the driveways of the plushest estates. You could spend hours searching for an acceptable shelter that's not already claimed. Once inside with the door barred, you'll still have to contend with the mess left by whoever won this shelter the night before. And, of course, the same will apply to your foraging area where you earn your bread during the day. The first human into the corner office will grab a big paycheck that day; latecomers will pick recyclables out of the garbage and take home pennies. This is not an efficient way to live. Most humans prefer to settle for territories that are imperfect but reliable. Many other animals feel likewise.

My territory consists of a shelter with some land around it. The shelter is a cute bungalow type, with a front porch designed to encourage social interaction with the neighboring humans. Built in 1917, when the local humans lived more simply and densely, it has three small bedrooms aloft, each with a tiny closet. Tacked on to the back is a box of new rooms, added to accommodate my newfound family and our modern notion of how much space—and stuff—each human requires.

The construction of the shelter is typical for settled humans, in that it is far more durable than humans ourselves. This building has housed more than four generations and was built half a century before I was born. That's not normal in the natural world. The mice who dwell in my yard excavate dens that are abandoned by the next year (often to be occupied by bumblebees). Wolves will sometimes dig one den and use it for a few years, but they're just as apt to dig a new one for each litter of pups. My neighborhood crows go without shelter entirely until they're ready to lay eggs each April, whereupon they build a fresh nest to protect the eggs and chicks. More like me are the burrowing mammals, like prairie dogs and marmots, whose shelters are used by one generation after another.

To make our shelters, humans will use whatever material is handy. In this regard, the species most like us might be the caddis fly and the bagworm. The larvae of these insects build mobile cocoons using anything they find lying about. A caddis fly case, constructed underwater, may be part stone house, part thatch hut, part twig shack, all held together with glue. Bagworm lodgings are built with the microlumber from dead plants. Similarly, the original portion of my shelter is a collection of materials that were abundant in 1917. The foundation is a stack of native rock, and the box above is made from local pine and fir trees. Humans the world over prefer to work with trees, if they're available. Stone is a common substitute, although you'll still need a few trees for the roof. Where the climate is too cold for trees, Inuit and other northerners build with snow and ice. On the American plains, too dry for forests, the aboriginal humans used leather stretched over a scanty skeleton of saplings. In the deepest Gobi Desert of eastern Asia, travelers have reported finding the abandoned skeletons of similar leather shelters, but in this case the skeleton was fossilized bones of dinosaurs. Of all the ingenious ways humans make shelter, some of my favorites are those dug into a strangely soft rock in central Turkey. In pinnacles and canyons, and in underground cities penetrating thirteen stories down, humans have carved stone into ant farms of shelter, connected with winding tunnels. Some of the pinnacles, which may poke seventy-five feet into the air, are still inhabited.

But back to my place. Surrounding my shelter is a patch of earth that I also call my own. As human territories go—or at least as they went until the dawn of agriculture, and as they still go among hunter-gatherer societies—it's laughably small, about one-fifth of an acre. Given that a family of seven crows requires several acres to support itself, my tenancy on a fraction of that looks precarious. Usually, the larger the animal, the larger a territory it needs to survive. A mouse might get by with a hundred square feet, but an elephant will need hundreds of square miles. A small band of hunter-gatherer humans usually needs a similar spread. If I were forced to rely on my one-fifth-acre territory for all my needs, it would be interesting to see whether I died of thirst before I ran out of squirrels and woodchucks to eat.

But my species invented agriculture, and this has revised human geography. Each human's need for an all-purpose territory has dwindled. Today, my food is produced on the territories of other humans, scattered all around the world. My water comes not from a pond in my yard, but from a lake in a town I rarely visit. The fiber for my clothing, the wood for my table, the clay for my dishes, all come from territories I have only an inkling of. If every inch of my own patch of earth vanished tomorrow, I would be just fine. Today, humans can survive on a territory as small as a park bench.

That's a novel phenomenon, if you consider the entire time line of human existence. At least, scientists presume so. For a glimpse of how the earliest hominids might have divided up their world, let's look to our fellow apes. Chimps, who are largely vegetarian, form groups of twenty to one hundred males and females, to hold a joint territory of about seven square miles. (Territories with second-rate resources can be ten times that big.) These members disperse across the territory to forage alone or with a few relatives. At nightfall, each ape weaves together branches to make a sleeping nest in the tree canopy. When chimps happen to converge in one area they'll cooperate to hunt colobus monkeys. But generally, small groups drift around in a core of the territory. The outer ring of the territory is visited only when a gang of males again works up the numbers and the nerve to conduct a patrol, risking battle with the neighboring group.

Gorillas, by contrast, are less territorial. Their groups are smaller and less rambunctious than chimps'. They'll drift less than a mile in a day's foraging. And while each gorilla group limits its exploits to a core area, these core areas often overlap with those of a neighboring group. When two groups stumble across each other, battle is rare. So, two of our close relatives present two different patterns of territoriality. And as I peruse the territorial configurations of even less-related creatures, I find a dazzling array of ways to slice the pie:

- In a migratory herd of one thousand Uganda kob (a fetching species of antelope), just a handful of males will win a "mating territory." These territories have been described as a cluster of golf greens in the middle of the savanna. Upon his green each champion will stand, glaring at his neighbor and awaiting the arrival of female customers. Females will not patronize a male who has not won a golf green.
- Among many terns and other colonially nesting birds, couples will build their nests exactly far enough apart that they can lean out and threaten to jab their neighbors—without risking actual contact.
- The sooty mangabey, a west African monkey, moves in family groups whose territories overlap. But to avoid unpleasantness, the dominant male of each group occasionally lets loose a "whoop-gobble." This spacing call allows other males to guide their families away, preventing a confrontation.

Many anthropologists turn to modern hunter-gatherer communities for clues as to how preagricultural humans might have spaced themselves. The comparison isn't fail-safe, but what is? And once again, the territorial behavior of various humans . . . varies. If I can form any sweeping generalization, it is thus:

In the beginning, bands of humans probably wandered like chimpanzees and gorillas, within a fixed territory. The degree to which they overlapped with, or clashed with, neighboring groups probably varied, depending partly on how scarce resources were. Within the territory, each human family, or at least females with offspring, might have staked out a private sleeping territory each night, or for a season at a time. Here are a few of the hunter-gatherer arrangements on which this guesstimate is based:

- The Dani tribesmen visited by author Peter Matthiessen in Irian Jaya defended distinct territories. A large (chimp-scale) group of these humans built quasi-permanent villages and practiced rudimentary agriculture. They considered some neighboring groups to be allies and others to be bitterest enemies. Deadly clashes and raids across the border were common. Within the territory, personal shelters were segregated by sex, with infants and girls sheltering with their mothers and boys sheltering with their fathers.
- Among the Bushmen of the Kalahari, a !Kung band of thirty humans might use a territory of three hundred to one thousand square miles. This may overlap the ranch of a settled farmer, but because the two cultures exploit different resources, there is no conflict. The territory may also overlap a neighboring !Kung precinct, but a tradition of sharing water holes and other rare resources precludes violence between neighboring bands. Personal territories consist of temporary straw huts used only for sleeping—one mated pair, with their offspring, per hut.
- The Mbuti Pygmies of the Democratic Republic of the Congo (formerly Zaire) inhabit territories shaped like the teeth on a comb, due to the dual nature of the resources they depend on. In the rainy season, each band migrates up its comb tooth to where a road forms the back of the comb. Here Bantu farmers have, since time immemorial, bartered with Pygmies for labor until the forest is productive again. Those farms at the base of each Mbuti band's territory are theirs to exploit. When the forest dries out, the Mbuti reverse their course and spend the rest of the year drifting down their comb tooth again. The Bantu farmers think the forest is scary, so they steer clear of Mbuti territory.

A few themes repeat themselves. One is that the typical subsistence culture perceives its territory, or aggregate territories, as sitting at the center of the world. This is understandable in any culture that lacks maps or a globe. Second, they often refer to themselves as "The People," "The Real People," "The People of This Land," or something similarly center-of-the-worldly. And finally, a consistent theme among them is that their identity is hooked to the spot on Earth where they were born. A young female who leaves her band to pair-bond with a distant male nonetheless continues to regard herself as belonging to her birthplace. And old humans will sometimes try to schedule their final wanderings so their dead bodies will rest in the places where they were born.

Although I can't say I think of my territory as the center of the world, the other two themes do hit home. While I live in South Portland today, I still think of myself as one of the "People of Boothbay, Maine," where I grew up. It doesn't matter that the isolated farm of my childhood holds no appeal to me. It anchors a corner of my identity, regardless. And certainly, the yearning to return "home" to die remains a common one. Although today fewer humans migrate "home" before they die, many make the migration after death, to be buried in their original territory.

Because most modern humans use much smaller territories, few now bury our dead near the shelter. As the division of labor within a culture splits tasks more and more finely, each individual territory serves fewer purposes. So it's no wonder my territory is a fraction of the size it would have been twenty thousand years ago. Like the crows, I could have more, but defending it would consume more time than I care to spend.

On the other hand, my territory has become more diffuse. Village commons and centers now provide a new form of shared territory held by all the members in a "band." Thus the cemetery up the street forms part of my property, too. As does the street itself, and the park downtown, and the beach two blocks east. For that matter, a diluted form of my territory extends clear across the state of Maine, which, like my birthplace, also clinches a corner of my identity. Maine, despite being an arbitrary line drawn on a map, somehow feels like my property. The entire political jigsaw puzzle of "the United States of America" holds some claim on me, too. Then my bond weakens as the boundary expands to include Canada; it frays further when I consider the southern Americas. Really stretching, my family's territorial history gives me a fingernail's grip on northwestern Europe. Beyond that, my territorial impulse wanes. Asia I cannot consider mine; Antarctica does not seem familiar. I wonder if the crows feel the same: that all of southern Maine, or even New England or North America, is homey, if not home.

BIGGER, HIGHER, WETTER: THAT'S BETTER

Mine is a safe and attractive territory, but it could be better. Gauging the quality of a modern human territory is a squishy business, because culture has so deeply infected the issue. For instance, in a herding culture, the ideal territory is a big one with lush grass. But in a fishing culture, humans would dismiss grass and scramble to control the best boat launch. And in my culture, which converts all its resources to money and uses that to advertise status, a territory's most important function is its ability to telegraph that status. If crows lived according to the rules of my culture, the couple owning the safest nesting tree would rank lower than the couple with a gold-plated nesting tree.

Because a human's status is now decoupled from natural resources, the symbols of territorial wealth are arbitrary. Thus, the standards by which my culture judges a territory can change with the wind of human fancy. One year, the top territory is a Caribbean island, but the next year it's a seat on a Russian rocket to the International Space Station. Perhaps the year after that it will be a castle in the Alps or an elevator ride to the earth's core. In most modern cultures, this is now the predominant means of claiming territory: competing to acquire whatever rare object is currently in vogue. If the Uganda kob had to play by these rules, then one year the stags would battle for a golf green, and the next year they'd knock heads over breeding spots aboard cabin cruisers. According to my culture, my territory isn't intended to provide food and shelter so much as to symbolize my rung on the economic ladder.

Shrunken though it is, and subject to the flickerings of fashion, my territory does still manage to meet some primal needs of yore. Even now, with humans so divorced from the land we inhabit, research suggests we still refer to a subconscious checklist when we assess a potential territory. The first theory to describe this primal hankering was the savanna hypothesis. It proposed that humans prefer anything reminiscent of the African savanna where humankind took shape. We crave a long view of an open plain, studded with fat antelope and a few trees we can clamber up when the lions appear. Research has shown that the preference for this landscape is indeed strong in children, although it fades in adults.

These days the savanna hypothesis is losing ground to the general evolutionary hypothesis. This theory posits that what we really want is much more: Long views are great, but we also want water, hills, places to hide, grass short enough to reveal the lions, and enough landscape variation to support a diversity of plants and animals that we can eat. That same variation also makes it easy for us to navigate by referring to landmarks. And finally, we want an element of mystery in our territories—rivers that curve out of sight, or hills blocking the view of . . . what? We retain an ancient desire, says the general evolutionary hypothesis, to be tempted into exploration. For a good example of this ideal landscape, you could visit the "melodramatic landscapes" section of your local art museum.

Few humans can maintain such elaborate territories these days. The cost of defending a savanna with hills, good climbing trees, and a curving river has risen considerably in the past fifty thousand years. But we seem to have boiled down that long list to a few affordable features. Most humans seem to agree: A long view is good, even if it's not your own territory you're gazing over. Water is good. Ample size is good. And green surroundings are better than brown ones. To see how these translate into modern territories, let us stroll around my neighborhood, and see where the nicest (most resource-costly) territories cluster.

Leaving my modest kingdom (very green and larger than average, but trees limit the view), let's head uphill. Here, at the summit of Meeting House Hill, stand some of the grandest of the town's old houses. The shelters are twice the size of mine. If it weren't for the urban forest that has sprung up since they were built, they would enjoy views clear to the horizon, with my distant friend Mount Washington looming in the west. And to the east (again minus the trees that have replaced open pasture) they would have gazed down to Casco Bay. What these territories captured was a high point in the landscape, which afforded them a view. (Each new shelter, however, received a compliment of young trees to provide that coveted greenery. The net effect, a few decades later, was a forest.)

Few features rate as high on a human's wish list as height itself, with its privileged perspective. Architectural history is written on the hilltops: Hill towns in Europe defend the highest point of ground. The Maya of Mexico built virtual hills of stone, while Mound Builder humans of the American South piled up soil that often elevated a human head above the surrounding forest. Marco Polo on his visit to Beijing encountered a royal hill built "a good hundred paces in height and a mile in compass," with a palace at its pinnacle.

Theorists explain our love for height as the instinct of a visual animal. We crave the earliest possible sighting of both potential food (perhaps a wandering herd of deer or the fruit ripening on a distant tree) and potential predators (a restless lion or an approaching army). Through the course of evolution, those humans who chose a camp with a view fared better than those who chose valleys, and so this preference became embedded as a biological impulse. It is not so different from the instinct that lures my crows to the tallest dead tree when they want to monitor their own territory. Or that prods seagulls to perch on chimneys, or wolves to seek a promontory from which to scan their empire or send forth a howl.

Modern culture has tinkered with this preference in a few ways. One is through high-rise buildings. Modern humans build our hills of steel and concrete, then divide them into myriad territories to rent or sell. And as a rule, since the invention of the elevator, humans have raced for the top floors. In hotels you'll find the most expensive suites at the top. In apartment buildings, too, the price rises at the summit. In 2006 *Forbes* magazine reported that humans will pay as much as 50 percent extra for ownership of the very top shelter, even if it's otherwise identical to those on lower floors. The competition over these rare territories can become extreme. The magazine located one penthouse apartment in Manhattan with a price tag of seventy million dollars—a value about thirty-five times more per square foot of shelter space than my rather squat shelter that sits thirty feet above sea level.

When I hunt for science on high-rise humans, the scant research I locate has been conducted mainly on college students in dormitories. But their reaction to their rooms does grant some insight. While students on lower floors tend to feel their rooms are too small and insufficiently private, students in identical rooms on high floors rate their rooms as light, quiet, and spacious. However, there does seem to be a price for climbing too high. A smattering of studies suggests that mental health suffers from living aloft. Children especially show more symptoms of mental illness when they perch in tall buildings. One study found that, for mental health, the ideal territory is a single-family shelter like mine. But in most of these studies, the scientists didn't try to determine which came first—the desire to live in a high-rise apartment or the neuroses of those particular humans.

Continuing our exploration, let's run back downhill past my shelter, all the way to the beach. Here at the shore is a patchwork of tiny old shacks and spanking-new mansions. In my settlement, as in wealthy settlements around the world, the hilltop is losing prestige to the waterfront. This is an interesting reversal. Not many decades ago, and starting at the dawn of time, humans have believed that waste dumped into lakes, rivers, or bays would go "away," vanish. Everything from human corpses to sewage and industrial chemicals went (and in poorer nations still goes) into these handy sewers. So as villages matured into cities, those humans who could afford to retreat from the stinking shores did so. In my sleepy town, the shore to this day is ringed with petroleum tanks, collapsed wharves, and parking lots. At low tide, discarded bottles and broken china from yesteryear stud the mud. But in developed countries, environmental regulations are speedily reversing the flight of humans from the water. These days in my town, when a human wants to erect a monument to her own potency, she heads downhill. She buys a sway-backed fisherman's house by the beach, knocks it over, and sets up a shelter with both a view *and* nearby water.

This thirst for water seems as hardwired as the hunger for a view. When an international team of researchers asked 432 humans from four different continents to rate a list of features they'd like (or hate) in a landscape, the loves and hates proved universal. Various water views claimed five of the top ten choices. Views of polluted water—a brown pool, a green swamp, a fish-littered beach—congealed at the bottom. These loves and hates taken together suggest that what humans love about water is its healthfulness. We are, after all, damp animals, who demand a daily drink. In fact, among the top water bodies on the list are nature's healthiest, waterfalls. (Fast-moving water prevents the breeding of mosquitoes and other pests, and falling water picks up oxygen that helps to purify it.) But I imagine our attraction to water also reflects our need to eat the other plants and animals that are likewise attracted to it. Humans are one of many predators who routinely stake out watering holes, letting thirst drive prey into range. (We're also in the company of many species regarding our suspicion of brown water, which among other hazards might hide a crocodile.) So my own territory could certainly be improved with the addition of a babbling brook or a waterfall.

A third major improvement to my territory would be an inflation of its size. In my current culture, a bigger shelter (and to some degree the territory around it) advertises greater economic strength. Mind you, my domicile is already grotesque when measured in the global standard of "humans per room." My mate, his two offspring, and I enjoy a ratio of about one-third of a human per room. This is about half the density in typical western European shelters, and one-fifth the average in Poland, where humans live at about 1.5 per room. The other common measurement is equally shocking: My shelter offers 600 square feet (56 square meters) per human. This is on a par with the average of Washington, D.C., which is the most space rich of about fifty international cities measured in one study. (Other shelter-rich settlements include Melbourne, Toronto, Oslo, and Stockholm, although D.C. is still about one-third roomier than these.) But now compare my shelter to those in poor nations. The Bangladeshis of Dhaka live in an average of 40 square feet (3.7 square meters) per human—that's smaller than a king-size bed. In Nairobi (Kenya), Dar es Salaam (Tanzania), and Antananarivo (Madagascar), humans squeeze into 55 square feet (5.1 meters) per human. At that rate, my shelter would be shared by twenty-six humans. I have indeed won a prize territory.

But still, it could be bigger. Let's turn down the coast, heading away from town and industry. When the road dips toward the ocean, the size of the shelters explodes. These dwarf even the showy new shelters built near my neighborhood beach. Here above the rocky shore and pocket beaches, high stone walls defend the boundary lines, and sub-shelters nestle under the trees. These monsters date back a century or more and were built to shelter a human family plus a few additional humans paid to maintain the territory and feed the family. Four of my shelters could fit in each one. The modern owners have converted many resources into money, and every human passing, by land or by sea, knows that at a glance.

Such big shelters and territories come at a cost. Whenever a territory is performing more than the basic functions, it is inefficient. The resident animal, be it a sparrow or a pride of lions, will be forced to spend more time and energy maintaining the territory. But every so often, an animal must thumb its nose at efficiency. The starling does not sing because singing is efficient. He warbles and tweets, gurgles and hoots, as loud and as long as his bank account of fat will allow. And if he gives his performance all he's got, then a great crowd of females may gather around his tree to admire the fabulous nest hole he has won. Having thrown economic caution to the wind, he will now have his choice of strong and aggressive mothers for his offspring.

Humans are more complicated, because our territories often serve symbolic functions in addition to providing shelter. My particular culture seems to be pulling us further and further from the biological mandate to mate and reproduce. Nearly every other creature (and plant, for that matter) spends nearly all its energy in pursuit of one goal: staying healthy enough to produce the maximum number of offspring. Not us. Not anymore. Although that goal remains central in some very poor cultures, the overwhelming trend is toward fewer offspring or even none. In this strange context, many human behaviors look nonsensical. Why, for instance, would a mated pair of humans battle for a rich territory if they have no plans to reproduce? There is no obvious biological answer for that. My best guess is that modern humans are a bit like domestic cats. Yes, the bag of cat chow has eliminated the need for a big territory rich in birds and mice. But the cat retains the instincts to defend a territory and to catch birds it's too sated to eat. I suspect the human, too, chases a family-sized territory out of antiquated instinct.

These grand old shelters are guarded by grand old trees. And beneath the giants lurk flowering rhododendrons, glossy hedges, and glamorous parties of flowers. These are large territories, and they're crammed with a diversity of plant life. This green wealth diminishes as we travel back toward town, and shrivels completely in the settlement center, where apartment dwellers resort to a few potted ferns on the windowsill.

LOUD AND CROWDED: TREACHEROUS TERRITORY

My shelter is modest compared to its shorefront neighbors. Stubby in stature and situated on a shallow hillside, it gives me a view that would never suffice in the event of charging elephants. And outside of the birdbath, my territory offers no water. But it could be worse.

It could be crowded with twenty-six humans, for starters. Crowding doesn't suit the human animal. Actually, it suits very few animals. Among many species, the rate of aggression rises with population density. Rabbits penned too thickly, even if food is plentiful, become more aggressive, breed less, and bear fewer offspring. Rats, too . . . but wait. *Male* rats become stressed by crowding. Females are fine with it. I've come to expect the two sexes to present divergent natural histories, and here again they do. It is males in the human animal, too, who seem to suffer most from crowding. The bulk of studies on crowding have been conducted on prisoners, refugees, and others who have lost control over their territory, which muddles the picture. Nonetheless, if you pack males into one jail and females into another, the males appear to suffer more health problems. Even if you pack males into a house, they get sicker, according to one study. Boys in crowded homes in India display a rise in blood pressure that girls do not experience. Just being in a crowd where they're forced to touch humans they'd rather not touch causes a male's stress indicators to rise.

But girls also suffer from living in crowded shelters when they're young. All human young tend to lose a sense of control over their lives. On average, they're less able to think for themselves, to ask for help, and to interact socially than are uncrowded young. The females especially suffer from "learned helplessness." This casts a sad light on the prospect of twenty-six Tanzanians sharing a shelter the size of mine.

On the other hand, isolated territories aren't ideal, either. (Isolation studies are mostly conducted on elderly humans, which again weakens their applicability.) Although all isolated elders appear somewhat at risk for negative moods and depression, this time females are the more vulnerable sex. Males aren't so pained to rattle around a territory alone. However, just living alone isn't enough to get a good female down: An Australian study found that a female human who doesn't share shelter is more likely to participate in a lively social network than is a solo male, and as a result she's less likely to complain of loneliness. (This was a rare investigation of humans age twenty-five to forty-four.)

If males don't suffer mentally from solitude, it's old news that they do go to pot physically without a female in the vicinity. The solitary male is sicker and dies younger than the mated male. This is likely a reflection of culture, not biology: Females in wealthy cultures push their mates to deal with minor health problems, preventing them from becoming major. Although the research on humans is sparse and sketchy, it's a reasonable bet that both genders of humans are similar to both genders of rats: Research on rats shows all of them are more apt to behave neurotically if they're housed alone than if they have a companion or two.

A crummy territory can result from many factors besides crowding or isolation. Humans don't like territories that are noisy, dimly lit, or that offer a view only of buildings or other human constructions. Humans in quiet shelters with plentiful light and a view of natural greenery report less stressful lives and a greater sense of well-being; patients in hospital rooms with a green view heal faster than those gazing on a landscape of bricks and cement. Insufficient control over the territory also causes stress. Humans who pay to borrow someone else's territory—renters—tend to have more health problems and shorter lives than those who have title to their territories, according to a Scottish survey, even after the effect of low income is factored out. Whatever the source of crumminess, all blighted territories seem to cause stress for their inhabitants, and stress in turn can cause any number of physical ailments. A bad territory is bad for you.

This list of territorial troubles adds up to bad news for humans who lack economic strength. Cabrini-Green, Chicago's infamous housing project, comes to mind. There was nothing prestigious about living in those particular high-rises. The upper floors have (or had—they're being torn down) views, but views were all they had. Many families felt like prisoners in their apartments, cowering under the assault of others' music and fearing violence when they ventured into the shared territory of halls and elevators. Groups of young males banded together and fought viciously for the right to sell drugs or rest on a particular patch of ground. The projects proved an excellent demonstration of how not to shelter humans who aren't able to maintain a territory unassisted.

Most animals will face an uphill trudge when their territory stinks. Male birds who can't win an attractive territory often find themselves standing on the stag line, while females flock to the landed gentry. Some female birds will opt to be the "second wife" to a male with a penthouse before they'll wed the owner of a fixer-upper. Although playing second fiddle will lower her odds of producing a healthy family, so will settling for a second-rate territory: Females who choose the fixer-upper also tend to raise fewer offspring. And the price of a poor territory can climb much higher than this. Male great horned owls who fail to stake a claim spend their lives sneaking from one territory to the next, poaching prey. They're locked out of the mating cycle entirely. These homeless "floaters" are also the first to go hungry when the snowshoe-hare population dives, and presumably they're the first to die of starvation. The same fate awaits an aging lion ejected from his pride by a young stud. And an even worse end looms before a homeless water boatman insect: He's eaten by his competitors.

THE INEVITABILITY OF INVASION

Why do humans invade one another's territories? After all, part of the reason territoriality works for animals is that it minimizes interaction and conflict between competitors. So why buck the system? Well, let's revisit my own territory. I think if we dash back in time, my territory will demonstrate a few of the myriad motivations that cause one animal to barge into another's domain.

In my neighborhood, territory grabbing may have begun as soon as glaciers retreated and chubby reindeer returned from the south.

Before written history, tribe after tribe of humans may have shoved one another off this land. The first documented shoving occurred as a flurry of European explorers crept near in boats to gauge the quality of land then held by the Armouchiquois tribe. After all, there's no sense invading if the territory is worthless. So the Europeans did as my crows do from time to time: They stepped just across the boundary line and took a look around. When my crows do this, they always (so far) conclude that the resources (worms and nesting trees) aren't worth warring over. They come home. But to Europeans, hemmed in at home by their own swelling numbers, this forested territory, beside an ocean a-thrash with fish, was irresistible. They went home and planned a proper invasion.

English settlers first put hammer to nail in the territory of the Armouchiquois in the early 1600s. We may never know if they asked politely for permission to camp or just made themselves at home. Different invaders use different approaches as dictated by their cultural ethics. Here in my ecosystem, English farmers eased slowly onto the shore. They did so with minimal conflict but under the same delusion that afflicted most of their fellow invaders: If no one human claimed a particular stretch of ground—as was the tradition back home—then the ground must be free for the taking. The Indians' view was that entire watersheds and mountain ranges constituted one shared territory, and *of course* no one human owned a particular square of it. This was somewhat beyond the Englishmen's ken.

Ignorance was bliss. Back in England, all the territories were claimed. Anyone who hoped to pile up more power by exploiting land had to first locate new land. And so investors sent farmers to my area to mark out new territories and set them to work producing livestock and vegetables. That's one reason to invade. These days, the majority of wealth-building territories are clearly marked and claimed, which makes this sort of land grab rare.

These days, the most common type of invasion in my neighborhood is the "sneak and snatch." This is a time-honored ploy among territorial animals who want to exploit a neighbor's territory without the trouble of actually taking it over. My crows practice this by tiptoeing across the territorial boundary and grabbing a couple of worms. They hop back home before they're detected. The beauty of this is that they preserve their own resources, and (hopefully) avoid a battle. It's the best of both worlds. Among the local humans, the sneak and snatch is typically practiced by young males. Like the crows, they sneak into territories on the sly, and rush off with jewelry and electronics. If the territory owner happens to discover them, they may put up a fight. But they would much prefer to flee back to the safety of their own shelters.

So far, my holdings have escaped such plunder. Invaders have tried to evaluate the assets of the territory, judging from the tracks they leave. Once, an intruder left the back gate open, and another time a sneak left tracks in the snow as he climbed the cellar door to peer in the window. Nothing in the larger territory enticed either of these raiders—not lawn mower, not ladder, not bird feeder. These resources would be too cumbersome to carry off in a quick grab.

My territory is somewhat protected by its surrounding territories, most of which are owned by humans with sufficient resources. For my neighbors and me, the trouble of invading far outweighs the benefits. We stay home. This is not the case in poorer areas. In one of my former neighborhoods, the benefits of invading often made the risks look piddling. The fact that invaders need to blend in makes an invader's own neighborhood his best place to raid. The reputed thief on my old block was a skinny male with ungroomed hair and clothing that looked too big and unwashed. His gait was tense and wary. He fit right in. But transport him to my current neighborhood and heads would turn. Everything about him would stand out against the background of healthy females ambling with baby carriages and mated pairs walking their dogs. Eyes would follow his progress down the street. If he turned up a driveway and disappeared into a backyard, someone would probably investigate. Because of this, sneak and snatchers tend to invade territories similar to their own.

Crime statistics reveal the pattern: Territories that are rented, not owned, are more vulnerable to invasion; so are crowded territories, and territories that appear unloved.

My shelter is frequently invaded, nonetheless. This morning our neighbor Hugh approached the shelter, stood outside, and drummed on the door. Although the door was unlocked, he waited until I opened it and stood aside. Then he entered. He had demonstrated that he would behave submissively as long as he was on my territory. That's the primary rule of peaceable invasion: If you're unarmed and agreeable, then you may be able to trespass without a fight. If you're sufficiently congenial, I might even volunteer to give you resources—a cup of sugar, a poker game, a bed for the night. My crows (a social species, like my own) also allow friendly visitors, especially if they're family. Crow kids who have grown up, married, and moved to a distant territory have been known to return home for weekend visits, much as humans do at the holidays.

And speaking of family, these constitute a curious category of invaders. Although a human family may share a territory and a shelter, there's little chance that they share all of it. In my shelter, for instance, my mate and I share only a few of the rooms equally. The others are bisected with invisible boundaries. The dining room table, for instance, is half mine. When my mate's piles of paper, keys, and books migrate onto my (neater) half, they're gently repatriated. My office is all mine, every heap of files, every box of books, all the dog-hair tumbleweeds: mine. He has his side of the bed, his half of the closet, and a vague entitlement to his son's bedroom. This is typical for a human family: Each member of the family claims sovereignty over some corner, whether it's a blanket on the floor of a hut or a suite inside a mansion.

Outside the shelter itself, my territory is much easier to invade. Although it's fenced to keep the dog in, a gate admits anyone who can reach the latch. This is typical of territorial design. The shelter, in my case, would be considered the "core area." It's the part most crucial to my survival. For my crows, the territorial core moves each spring, depending on where they construct their nest. Their core will also contain the best worming areas and perhaps a water source. The outer reaches of the crows' territory, and mine, may hold resources, but they also serve as a buffer. This is the part of my territory into which a neglected neighbor boy throws his baseball repeatedly, giving himself an excuse to crawl through the hedge. It's the part of my territory in which another neighbor allows her cat to roam and kill birds. It's the part—the airy part, anyway—that the village bully (every settlement has one) invades with loud music on a summer day.

As the rings of my territory expand into the public spheres, they're easier and easier to invade. Just the other day I found myself displaced at "my" fish market in town. Those of us who normally use this public territory park our cars in the spaces painted out for that purpose. But as I pulled up, a large and expensive car with out-of-state plates sailed in and parked so as to block three of the painted spaces as well as the front door. By the time I entered, three post-breeding-age females in shiny attire were testing the defenses of the two young males who own the store. "I bet they don't even know how to fish," one female cawed loudly to the next. The males tolerated the aggression until they had extracted resources from the invaders, then they barked a hostile "Thank you!" at the retreating sequins. I, rather than confront the invaders, waited for them to leave, knowing I'd soon have this public portion of my territory back without a fight.

THE OBLIGATION OF DEFENSE

For as long as humans have claimed territories, we've been faced with the need to defend them from trespassers. If the whole territorial paradigm is to pay off, then competitors must believe that crossing the boundary will have dire consequences. The most proactive defense is to make the boundary obvious. Among tigers, this is done by scratching trees and scent marking boundaries with urine, anal-gland secretions, and feces. My crows and many other birds light at various points in their domain to voice a special territorial clamor. A beaver carries mud from a stream to erect a chain of mounds at the perimeter of his tract. And humans? The variety of boundary markers we deploy defies enumeration.

I subscribe to the typical Western pattern. A stranger approaching my territory would note first a hedge growing between the public sidewalk and my shelter. The invader would note a path branching off from the sidewalk and passing through a gap in the hedge. Then steps ascending to a roofed porch. And finally a door. The nearer he got to the core area, the more solid the defenses would appear. As I approached this stranger from inside, I could assess him through the glass door before deciding whether to release the lock. As humans have done for thousands of years, I also keep a dog who alerts me to the approach of both friend and foe.

But mine are thin defenses. When I lived in New York City, my sidewalk bulwarks were no flimsy hedge but a wrought-iron fence. Then two separate doors blocked entry to anyone without a key. Inside, my second-floor apartment bore still more locks. My friends on the lowest level of the building proclaimed their strength with wrought-iron bars across their windows and a grate across the door. After a neighbor couple lost entire shrubs from their garden to sneak and snatchers, they chained the replacements to the fence, signaling their intention to defend what they owned.

And now even those are outdated defenses. The latest in human boundary markers evokes the earliest human settlements, the towns that sprung up with the advent of agriculture. The gated community of today is a cluster of shelters ringed with a high wall to foil invaders. Portals to these settlements are few and guarded either by a human or electronic sentry. These territories are rarely invaded.

Yet to come, perhaps, is a reversion to the old "pull up the ladder" defense. The bastel houses of northern England weren't cheery, with stone walls three feet thick, and arrow slits for windows. The ground-floor stables were connected to the upper-floor dwelling by only a ladder. When invasion was imminent, the territory owners ascended and pulled the ladder up with them. Ladder defense was practiced all over the world. In Ihlara, Turkey, I once found myself in a long canyon whose walls were punctuated with black doorways and windows, all located ten or twenty feet off the ground. Behind those doors, I knew there were ancient churches, storerooms, and housing carved into the soft rock. And maybe even old ladders. But I couldn't get there. From the perspective of a would-be invader, I found the arrangement perfectly frustrating. Cogitating on this, I now think perhaps the same effect is achieved in a high-rise building whose elevator requires a key. Neither ladder nor elevator is foolproof, but both are probably an improvement over an apartment on the corner of a city block, which, judging from crime statistics, is the most attractive location for sneak-and-snatch invaders.

Now, I've never experienced a hostile invasion of my shelter, so I don't know how aggressive my defense might be. I do know that I'm plenty irked by invasions of my larger territory. I'm only mildly irked when the boy next door crashes through the hedge. What I mind vehemently is other humans letting their invasive weeds and their murderous cats exploit my territory.

This raises a curious issue: Almost invariably, animals defend a territory only against their own species. My crows, for instance, couldn't care less about all the goldfinches, squirrels, and woodchucks in my yard. And ditto for the goldfinches—they'll only skirmish with other goldfinches, leaving the chickadees and house finches in peace. And likewise with humans, to some extent. So I happily share my territory with the animals and plants native to my realm. In fact, I receive an aesthetic and moral benefit from this native community. But my benefit is threatened by cats. And English sparrows, European starlings, Japanese bamboo and Oriental bittersweet vines. All compete with the natives for the plot of ground we call home. And so I defend against them—with weed whackers, clippers, and whatever harassment comes to hand at the moment of the invasion.

If a hostile human did penetrate my defenses and enter my shelter, how might I react then? One can't know until one has faced an invader. It's a good bet, though, that I'd react like other territorial animals. The "flight distance hierarchy" describes the series of actions a defending animal will take, depending on how close the invader is. It's a pretty commonsensical formula. Imagine that you have discovered an invader in your shelter. When the invader is a good distance away, the most likely response—for you and other species—is to flee. If the invader is close enough that your flight would attract his attention, you might opt to hide (a "freeze" response is more common in other species). Closer still, and your options will narrow. You'll be forced to confront the invader and issue a defensive threat. A cornered raccoon would hiss and growl; you might scream or grab a kitchen knife. Finally, if the invader is close enough to render all options moot, you'll attack like the proverbial cornered beast. But that's the least appealing option, and the riskiest, for any species. Given a chance to escape, most will jump at it. Lionesses, when they detect an intruding pride, do something I find ingenious. They listen carefully, counting the number of roars. If the "away team" is smaller, the home team will rouse themselves and rout them. If it's bigger, it may be allowed to trespass—unless resources are particularly sparse.

Human females, especially, strive to avoid a fight. In a recent study of the two sexes' response to various threats, the difference was stark. Females were much more likely to flee or hide at the first sign of trouble. Given the same sign—a stranger jumping from behind a bush or who turns surly after bumping into you—males were more inclined to investigate the situation or seek a weapon. Overall, females were much quicker to shout for help, a response that males denied resorting to under any circumstance. I know that in my own instances of great fear, the instinct to communicate my peril was so strong that the screeches rushed from my throat before I had any notion of summoning them.

Happily, not every attempted invasion leads to such a showdown. Usually, two competitors have a chance to size each other up before committing to an invasion or defense. If I look out my door and spy a peddler of life insurance or religion, my frown or hand flap is usually aggression enough to rebuff the invader. If the invader is unfamiliar, I might investigate his motivations through the glass. If he seems menacing, I'll retreat to my core area and spare myself a fight.

In the rare event that push does come to shove, the famous home-field advantage can be a real asset. When biologists first noticed it among wild animals, they did the scientific equivalent of rubbing their eyes in disbelief. Yet it held up under closer scrutiny: a sniveling runt of a bird can prevail over a big bruiser if the showdown takes place on the runt's territory. This has been replicated in a number of bird species, and in humans, too. The human version is best studied in the sports arena, where folk wisdom has long maintained that the hometown team always has an edge. And careful analysis shows this is, in fact, the case. A huge study of four hundred thousand games—football, baseball, hockey, soccer, basketball—played in England and the United States over a span of 127 years showed that the effect is strongest in the earliest years of a sport. Then it tends to fade. But other studies show that it also varies considerably among countries, and even among teams from one year to the next. The researchers concluded that the home-team advantage cannot be attributed solely to the support of the spectators, as was originally thought. Rather, it appears that the home team's familiarity with the field and the away team's travel fatigue may combine with the biased crowd to tilt the playing field. Human territory raiders also illustrate the home-field advantage. The majority of them will bolt for safety if the territory owner discovers them.

Among other animals, home-field advantage probably results from factors similar to those that tilt human sports. The resident chickadee in my yard will be more familiar with the ambush spots and hazards of his own territory. He may be better rested than an interloper with no reliable source of food and shelter. And he could also be riding a testosterone high, which many animals experience after winning a fight. In fact, experiments have shown that testosterone alone can make or break a bird's territorial defense. Male owls who were given testosterone implants hollered bloody murder at an experimenter's loudspeaker when that loudspeaker played another owl's territory call. And the male Scottish red grouse, when similarly implanted, defended such big territories that the season's youngest males were utterly frozen out. The winners got such a high from their success that they did just as well the next year, even without implants. The losers, without territories, faced starvation. To further stack the deck in favor of the occupant, an invading bird may be riding a testosterone bummer if he's already lost one fight over territory. A loser's testosterone level often crashes, resulting in a shy and gloomy bird.

Just what chemistry causes a female like me to act more or less territorial isn't known. Laboratory experiments with mouse and rat mothers suggest that nursing her offspring causes chemical changes in a mother's brain. So perhaps those changes allow breeding females to put aside their fear when their offspring need a champion. But many females are territorial even without offspring to inspire them. The famously butch hyena females take the lead in marking the clan's territory, patrolling it against trespassers, and launching strikes against neighbors. In monogamous klipspringer antelopes (Africa), it's the female of the pair who attends to the boundary-marking chore, while her mate follows behind attempting to cover her marks with his own. Female tigers mark their territories and defend them against other females. Even the Oriental fruit fly will guard a peach on which she lays her eggs, head butting other females who attempt to invade.

In the human, it's rare for females to rip into an invader if a male ally is willing to do the job. This is also true of chimpanzees. It is males who make the silent, single-file journey to the boundary of their territories, hoping to surprise interlopers. Female chimps rarely participate, and never during the fertile phases of their cycles. The only case I've encountered where human females habitually take the frontline is among a Pakistani gypsy-style culture. When police attempt to break up a camp, it's the females who charge from their tents to confront them, while males and offspring scuttle out the back. As diverse as human cultures are, I'm surprised to find only one example of such role reversal.

In addition to defending my core area and yard, I also maintain a bubble of space that surrounds me wherever I go. I feel it bump up against competitors when I stand in a long line at the grocery store, or when another human at the gym takes longer than her allotted time on the machine I want. My bubble shudders with irritation when another driver cuts too closely in front of me or follows too near behind. In fact, one of the seminal studies of human territoriality focused on drivers. In 1997, a sociologist measured the time it takes for a driver at a shopping mall to vacate a parking space. He found that a human takes seven seconds longer when another human is waiting for the space. And if the waiting human blasts his horn, the occupant will defend the temporary territory for an additional twelve seconds. This shocked me, as a human who often hustles to accommodate my fellow humans. Even more shocking was this detail, lost in the headlines: Males (and only males) will actually abandon a territory *quicker* if the intruder is driving a more expensive car. Females are either unimpressed by such status displays or, like me, they don't know their Alfa from their Edsel.

Males and females defend their personal space differently in other contexts, too. While both sexes seem to agree on which territorial invasions are risky, they disagree on the degree of risk. Females are quicker to flee public spaces, especially those that are poorly lit or devoid of other humans. Females are also more suspicious of strangers who approach them in public territories. None of this surprises me, given the female's smaller size and the male's greater propensity for aggression. Territorial invasions usually will be more dangerous for females than for males. For females, it is especially risky to defend a territory when they could flee instead.

THE CURIOUS CASE OF NOMADS

It occurred to me during a visit to Sami reindeer herders in Norway that nomads are anything but homeless. Rather than wander aimlessly, the herders were following the equivalent of green highways connecting their winter homes to their summer homes. Prehistorically, the aimless-wandering version of nomadism probably waxed and waned in concert with the human population. In places where a plague or catastrophe wiped humans off a swath of land, other humans might wander with impunity. But wherever the planet was populated, surely it was also defended. So it is with chimpanzees and other territorial animals: If you take to wandering, you'll soon find yourself in someone else's backyard. And even if you do manage to wander far without trespassing, you're going to end up in an ecosystem whose food and shelter resources are a mystery to you.

So the humans we call nomads are more like retirees who own a summer territory in London and a winter territory in Miami. They beat a regular path between their resource patches. The nomadic Sami I encountered were migrating from inland winter camps (villages, these days) to the coast for summer grazing. Before modern medicine swelled the human population, the Sami may have rushed for the coast in a free-for-all, claiming summer territories on a first-come, first-served basis. That appears to be a common arrangement when a shared resource (grazing land) is plentiful. But equally common is the phenomenon of populations swelling beyond the capacity of the resource. And when that happens, the usual human pattern is to split up what was once a shared resource. Patches of land may be allotted to a clan, to a family, or to an individual. They may be unmarked, or they may be bounded by streams, ridges, and other natural features. But they are owned and defended. Moving with the seasons, the nomads travel public paths that connect their two territories.

Most interesting of all the nomads are the gypsy cultures. (I use the term "gypsy" loosely, for humans who travel through a territory, making their living from other humans.) Circulating in Pakistan, Afghanistan, and northern India are tribes of humans whose main resource is neither pasture nor hunting ground, but the villages of other humans. According to the Pukiwas culture in Pakistan, each family claims not just a physical territory, which encompasses a circuit of villages, but also a professional territory—a set of skills and services to offer within that circuit. To my wonder, I read that these professional territories have been so fiercely defended over the eons that each "trade group" has come to speak a distinct language. The Chungar, who make baskets and brooms, speak a different language from the Kanjar, who specialize in terra-cotta toys, carnival rides, music, and prostitution. And the Qalandar, who train animals and perform magic tricks, speak a different tongue from the Lohar, who are metalsmiths.

Each profession uses a different resource in a village (the Lohar exploit the "farm implements budget," while the Kanjar work the "entertainment budget," for instance). Therefore, their territories can overlap without friction—as a crow's overlaps a hummingbird's. But two jugglers in one neighborhood? That's untenable. So the Pukiwas have developed a method for marking a temporary territory, be it a street corner or a market square. From time to time each Pukiwas who is working a territory will produce a distinct sound. She may hoot or whoop, strike a drum, or sound a note on an instrument. The Pukiwas are said to know several hundred humans by their territorial call alone. If the invader is offering the same goods or services as the territory holder, she'll detour to unexploited ground.

Among those Pukiwas who peddle their goods from house to house, another territorial tradition holds sway. A female who sells brooms and baskets "owns" certain clients. Their shelters constitute her territory, and woe betide the competitor who should seek to invade. As with many other animals, a Pukiwas mother's territory becomes the territory of her daughters, in time. Among the Pukiwas, these human territories are far more crucial and enduring than their tent camps at the edge of town.

■ ■ ■

Until recently in my own culture, human territories passed, like the Pukiwas', from generation to generation. A farm established by the parents passed into the ownership of the offspring, and then the next generation, and the next. This still happens from time to time—my cousin now owns the shelter our grandparents built and inhabited.

But territory has become more symbol than substance for many humans. I love the shapes and colors of my shelter and the greenery of my yard, but I'm not dependent on the services they provide. I could as easily adopt a territory down the street, or halfway around the world. Only in the wilderness beyond the reach of refrigerated trucks would I start to worry over the fertility of my soil and the clarity of my waterfall. And so my territory is a luxury and an advertisement of my status. Because my food and future don't depend on it, I won't defend it with any particular vigor.

Nonetheless, I retain an ancient human preference for a territory that could support me entirely. When I traded territories ten years ago, the hardest adjustment was to the limited view here. My previous shelter looked out from a hillside over a mucky cove, a freeway, and what I fondly called the Scrap-Metal Mountains. But it was a view. I miss it still. And I can't say I'd object if my shelter magically transported itself to the oceanside, either. On the other hand, I control far more trees and shrubs here, and the noise of that old freeway is a distant purr. It's an adequate, wholesome holding. And more than spacious enough for its occupants. Many humans around the world do, in fact, survive with a territory as small as a park bench.

6
HUNGRY AS A WOLF:
DIET

The consummate omnivore, *Homo sapiens* can derive nourishment from some thirty thousand plants, as well as animals ranging from insects to birds, fish, crustaceans, reptiles, amphibians, fellow mammals, and even Hostess Twinkies. It can survive on diets ranging from vegetarian to nearly all animal. Accordingly, the dentition, a combination of slicing and grinding teeth, is reminiscent of other omnivores, such as black bears, wild pigs, chimpanzees, and skunks.

As with any species, insufficient food does act as a limiting pressure on population. The starvation rate for this animal is lower than for many others. Nonetheless, of all the human young that perish each year (twelve million), the failure to find food is the underlying cause for about half the deaths.

Like a few other species—termites and leaf-cutter ants, primarily—the human processes much of its food rather than consuming it fresh. This began with the human's control of fire, to which the animal now subjects the majority of its foods. But the human also treats food with fungi, acids, salt, bacteria, and alcohol to preserve it for consumption in the future. Like chipmunks and ravens, humans will cache food, retrieving it days or even years later.

Homo sapiens has some capacity for storing food energy within the body, like a bear or a camel. But this capability, when combined with the animal's instinctive craving for high-calorie foods and its newfound ability to produce vast amounts of such foods, now causes considerable morbidity. The least disabling symptoms are discomfort and impaired mobility, but joint failure, circulatory disease, and metabolic disorders can result in death. The foraging style of many humans now borders on suicidal, an anomaly in the natural world. Regardless, it's spreading like butter on a hot English muffin.

STEAK OR POTATOES?

What am I supposed to eat? I mean, what would a *natural* human *naturally* eat? Fruits and vegetables? Cows and chickens? A little of each? A lot of each? My biologically mandated diet is the subject of a big fat food fight among scientists. The factions argue their cases by pointing to the diets of various relatives and asserting that we should eat whatever they eat or ate. These relatives include the gorilla, the chimpanzee, the Neanderthal, and hunter-gatherers who persist in a few wild corners of the planet. Each of these relatives observes a distinctly different dietary practice. As a result, a reasonable human could get the impression that we're supposed to eat only leaves 'n' stems with insects on them. Or almost all meat. Or a chimpanzee's diet of fruits 'n' monkeys. Even within the hunter-gatherers, diets differ immensely. After all, Inuit are awash in caribou and seals, while the Bushmen of the Kalahari Desert are not. The Bushmen dig in the ground for fleshy tubers, but this would not be a profitable enterprise for the Inuit.

It's a sorry excuse for an animal who even has to ask, "What should I eat?" Bears don't trouble themselves with such introspections.

Among humans the situation has become so nutty that one contingent cries out that the human should eat exactly what its body *wants* it to eat. I love the idea, on a purely hedonistic level, but practically, we'd have to renovate all the doorways in the house to accommodate the butt I would develop under that policy. My body wants to eat everything, all day long—as long as it's not sour. My stepdaughter would present a different problem—protein does not remotely interest her body. Her brain reminds her from time to time that she ought to have a glass of milk.

But since the question has been put, let's take a sober look at what our various relatives eat (or ate), and see if any of the menus inspire salivation.

Because gorillas, chimpanzees, orangutans, and bonobos are our closest animal relatives, some humans have decided we should follow their dietary examples. Now, it should be noted that the folks in question can seem a bit unsober. One advocate of eating all foods raw has gone so far as to declare, as an argument against drinking milk, "Nobody has ever seen a gorilla milk a buffalo in a primeval forest." This prompts the question of whether anyone has seen a gorilla wear a Rolex in the primeval forest, or discuss supply-side economics in the primeval forest. It's what gorillas *do* eat in the primeval forest that interests me, and that information is easily obtained.

Gorillas eat leaves, stems, pith, fruit, bark, bulbs, roots, shoots, and vines—all raw. They consume only a smattering of animal protein, in the form of insects who find themselves on the wrong leaf at the wrong time. Another great ape, the orangutan, is also inclined toward the veggie side of the menu, although stray reports describe an occasional female catching such exotica as a slow loris (a doddering primate), a rat, and a gibbon (a lesser ape). And ditto for the chimplike bonobo, although this great ape more routinely snarfles easy protein like worms, beetles, and the occasional small mammal.

The chimpanzee, our very closest animal relative, is different. Chimps eat more fruit than their fellow apes. They also use tools to open high-calorie coula nuts. And they eat meat. With a coordinated attack worthy of any social hunter (female lions, coed humans, coed wolves) male chimps pursue and tackle everything from midsize monkeys to bushbuck, wild pigs, small rodents, and mini-deer. Like their fellow apes, they do not subscribe to cooking. So, in a coula nutshell, my great-ape relatives eat raw veggies with insect croutons. On holidays, some species will serve a raw monkey.

Right away, I'm dubious that this is the diet I'm meant to eat. Very few studies have analyzed the health of humans who eat a raw-food diet, but the ones I've seen are worrisome. The longer humans participate, the skinnier they get, to the point that three out of ten females studied ceased to menstruate—a sign that their bodies could not support reproduction. Another study found that raw-food vegetarians suffer low bone mass. A third discovered that serious raw-food eaters, who consume an average of twenty pounds of fruit a week, showed significantly more tooth erosion than others. So, do let's consider some other eating plans!

What sustenance fueled our hominid ancestors? The first humanlike animals to part ways with the ape lineage were the 'piths—*Ardipithecus* and the australopiths, like the famous Lucy. But what they ate three to five millions years ago is a mystery. Without fossilized platters of food, dietary estimates depend rather heavily on the analysis of teeth. This analysis takes place both in the visible world and in the microscopic realm.

The macroview available to me in the mirror reveals a variety of tooth shapes. My incisors are like shovels. The canines are like spears. And the molars, taken as a group, bring to mind the pyramids on my kitchen meat mallet. At this scale, teeth are self-explanatory. My incisors are serrated knives, good for slicing and scraping. The canines are for stabbing. (The reason my canines are so dull is that humans have cut way back on biting our enemies since we learned how to make an external spear. Chimps, who are big on biting, still have big spears in their mouths.) Finally, the molars are like interlocking meat mallets, designed for crushing and grinding. Mine is not a textbook carnivore mouth. In the mouth of a meat-eating weasel or wolf, pointy shears line the jaw like a parade of sailboats. They're ideal for slicing meat into chunks that will fit down the gullet with no further chewing. Nor have I the standard herbivore design. In the jaw of a leaf-eating horse or elephant you'll find flat-topped teeth with ridges like old-fashioned washboards to grind vegetation. What I have is knives in the front, for slicing off bites of food, and molars that then grind food up. The molar cusps are sharp enough to do some cutting and dull enough to do some mashing. This is the kind of incisor-molar mix that lines the mouth of omnivores such as pigs, chimpanzees, and black bears. So, to get back to the question of what the human ancestors ate, what sort of teeth was my ancestor Lucy wielding?

Judging from the fossils, the incisors of her species were a bit more delicate than the broad shovels a chimp uses to scrape plant tissue. Her canines were inoffensive things like mine. And her molars, although less peaked than my own, were nonetheless an omnivore's. However, this does not prove that Lucy ate antelopes on the plains of Africa three million years ago. Anthropologists are skittish about leaping to such conclusions. Even though most of the great apes eat some meat, and modern hunter-gatherers eat meat, still a great many scientists will not declare Lucy a true omnivore until a new *Australopithecus afarensis* is discovered with a fossilized sparerib in her mouth.

. . . Or until microscopic analysis finds teeny meat strings lodged in tiny pits of single teeth. Peter Ungar is an American tooth scholar who studies the tiny scratches that food leaves on your teeth, my teeth, and Lucy's teeth. Different foods leave different traces. Fruit causes pitting, for instance, while leaves make streaks. And when Ungar looks at Lucy's teeth he sees pits, streaks, and microflakes, indicating a lifetime of chewing fruit, nuts, seeds, and tubers. So according to tooth marks, Lucy ate a wider variety of plant foods than a modern chimpanzee, and fewer cows than a modern Texan.

. . . Or until isotopic testing reveals teeth built from pure meat isotopes. This test takes advantage of the fact that ancient hominids and grazing animals did not share a taste in plants. Just like modern cows, early cow-type species ate so-called C4 plants—grasses, mainly. Early hominids, however, browsed like the chimpanzee on C3 plants—fruiting trees, broad leaves, and roots. Each plant's isotopic signal was stored in the tooth enamel and bones of he who consumed it. So the bones of ancient hominids should be carrying only C3 isotopes, if they were vegetarians. But if they ate grazing animals, they would have inherited the C4 stored in those creatures. So when a team of archaeologists tested the teeth of one of Lucy's descendants, lo and behold: Three million years ago either *Australopithecus africanus* had developed a strange fancy for grass, or she was eating grazers.

Thus, on the strength of tiny scratches and isotopes in antique teeth, the debate over meat versus veggies wobbles through the ages until about two million years ago. By then, at the very latest, a full-blown *Homo* was definitely biting into a meaty lifestyle. Along with crows, skunks, wolves, wild boar, and a host of others, hominids developed a dietary flexibility that would allow them—and eventually us—to thrive in oodles of ecosystems.

WHATEVER THAT IS, COOK MY PIECE

If our dietary range isn't unique, our concern with temperature is. The human animal likes its food hot. When I think of eating meat, I think of a rib-eye steak blackened on the outside and a glistening shade of rose within. That version may not have landed on the serving leaf until about one million years ago (archaeologists quarrel about charred evidence that may go back two million years, or only a quarter million). *Homo erectus* was the first *Homo* who all archaeologists concur was a barbecuer: He not only ate meat, he ate it hot. Before that, hominids ate most or all meat au naturel, using their teeth or stone cutlery to tear raw muscle loose from the bone and sinew. If you're under the impression that cooking makes meat tougher, then you've never tried to eat a raw steak. I have, just to see for myself. It slithers between the teeth, crushing a little, but then squirting free. To reduce it to pulp demands minutes, not seconds, of chewing.

And that was the flesh of a sedentary cow. Wild animals, who exercise their muscles all day, are tougher. The white-tailed deer Mom shot one fall when I was a kid comes to mind: The meat was dark, pungent, and as tender as automotive rubber. And that was cooked. Had I tried it raw, I expect I'd still have shreds of Mom's deer between my teeth today.

The spot on the Time Line of Prehistory where a cooking fire glows marks our family's departure from the dietary path of all other animals. Prior to *H. erectus*, my ancestors probably used fire only to scare off leopards and other nighttime predators. But the nutritional options of *H. erectus* multiplied on the day he dipped an antelope shank into the leopard repellent.

Fire made such a difference in our diet, and hence our brain, that one primate expert from Harvard has deemed our species "the Cooking Ape." Dr. Richard Wrangham talks the talk, but he has also walked the walk and chewed the fewd: He once ate a chimpanzee diet of raw, fibrous forage for a week. "Chewing raw food," reported Wrangham, whose deep-set eyes and wide grin lend him a chimpy impishness, "requires a lot of work." Chimps in the wild spend five hours a day chewing to break down the plant cells that contain nutrients. The cooking fire was humankind's first time-saving kitchen appliance.

Among fire's other advantages, it softens food. Even the toughest wild meat becomes easier to slice after cooking. In raw meat, the intact muscle cells resist puncture. A determined pair of molars can mash the muscle but can't easily slice it or free the juicy contents of each cell. As heat penetrates, meat proteins harden. Now the sharp edges of a tooth can shear through the cells, simultaneously reducing the size of the morsel and bursting the juice balloons. Yum. Of course, over-roasting meat causes those muscle cells to shrink too much, squeezing out all moisture and producing a dental hazard. Fortunately for a *Homo* new to the fire appliance, this error is not irreversible. A second round of chewability arrives when the connective tissue between the dried-out cells reaches its own limit and dissolves into a slippery marinade: That produces the "fork-tender" texture of long-barbecued meat.

Now, anyone who has worked in the restaurant business knows the primal pull exerted by an open flame or even a vat of hot oil. Once a *Homo* knows that both onion rings and potato sticks are much improved by a dip in the Frialator, then it's just *Homo* nature to dunk battered peanut butter sandwiches, maraschino cherries, and dill pickles. And thus did the ancestors discover a world of heated cuisine. Or it could have happened the other way around—first with roasted nuts, then on to meat. Either way, they took to cooking like a duck takes to water.

Lest we start to take humans too seriously, processing food is not a sign of staggering genius, judging by the other animals who do it. Various species of leaf-cutter ants carry snippets of leaves down into their nests, where they inoculate them with a special mold. They give the mold time to break down the leaves, then they eat the mold. When a young queen leaves home to start her own nest, she takes along some fungal "seed" for the new farm. Similarly, hundreds of termite species (who aren't ants, despite similarities) also use fungi to transform plant fragments into processed food. A species of damselfish tends algae gardens to sustain itself, and an aquatic snail appears to rip holes in sea grass, then fertilize the gash with his manure, producing his own favorite fungus. Eating is straightforward for most, but not all.

No, the genius of *H. erectus* lay not in processing food. It was in using fire to multiply the number of foods he could eat. If you can extract nutriment from hundreds of plants and animals, then you're better adapted to contend with change. Ask the panda, whose devotion to bamboo shackles him to a single ecosystem, which is shrinking. Or inquire of the koala, who along with his eucalyptus tree is confined to isolated patches. No such single-mindedness hobbled my forebears. Confronted with hard tubers, *Homo*s softened them in fire. Tough-shelled nuts? Let the fire pop 'em open. The fire also made feeding more efficient. Legume seeds (beans, chickpeas, lentils) that caused cramps when eaten raw yielded more nutrients and less pain after cooking. The *Trichinella* worm, who once passed easily from raw boar meat to living hominid meat, died when it made a detour through fire, leaving more nutrients for the *Homo*. Ditto with a plague of other parasites. Some starchy vegetables, when heated, were quicker to spill their energy. In others, heat worked upon vitamins—modern summer squash and carrots offer more vitamin A when cooked than when raw. Moldy seeds, once cleansed by fire, were less likely to cause kidney failure. These improved foods freed a *Homo* from a constant hunt for low-grade calories. Those hominids with culinary leanings endured fewer aches and fevers, cramps and tumors. They thrived at the expense of those who continued to eat raw.

Physiological side effects happened, too. Bones and teeth are expensive to build, and evolution rewards economy. So when the human jaw was no longer required to chew for hours a day, those individuals born with a lighter jaw had more energy to spend on other things. Over eons, the hominid jaw daintified. The teeth, too, diminished. Even the gut shrank, as fire did some of the digesting ahead of time. Whereas apes tend toward a bulging belly, the hominids sucked theirs in. Exploiting the energy saved elsewhere, the body may have grown a bigger brain. And maybe the myriad mutagens that form when food is cooked caused our DNA to mutate a little more frequently and nudged up our rate of evolution. Maybe.

That's generally the way diet evolved, as the *Homo*s evolved toward *sapiens*. The stone cutlery improved, a stone grinding appliance joined the fire in the kitchen, and the undertaxed teeth and jaw continued to shrink.

Is fire the reason my jaw is short four teeth today? The normal complement for a human is thirty-two—eight slicers, twenty break-and-grinders, and two rather ceremonial canines. I'm missing the four rear grinders, or wisdom teeth. The bottom two made a belated entrance in my midtwenties. They failed to negotiate the curves in my diminished jaw and were torn out by a dentist. The upper two simply never existed. I take some pride in this indication that I'm on the cutting edge of evolution. The occurrence of wisdom teeth varies enormously at this point in human history. A Czech team investigating the remains of prehistoric Tasmanians (now extinct) found the wisdom teeth there intact, whereas prehistoric Mexicans had entirely washed their hands of third molars. Even with our modern gene mixing, racial differences persist. Another study in Mexico found one in four modern citizens is shy at least one wisdom tooth. But a survey in Croatia found just one in twenty down a tooth. And nearly three out of ten Chinese humans in Singapore are missing one or more. The truth is, fewer and fewer humans have room in the shrinking mouth for that final set of molars. This problem has plagued our species long enough that the risks—infection, impaired chewing, even tumors that form around impacted teeth—have eliminated some thirty-two-toothers from the gene pool.

Before we look at modern human diets, one more relative—scientists aren't sure where he fits in this story—deserves consideration. Back when the fresh-faced *H. sapiens* poked her Africa-born head into Europe, she encountered a handsome cousin, *H. neanderthalensis.* Neanderthal had been enjoying Continental cuisine in peace for about 150,000 years before this upstart arrived. He had lovely kitchenwares, and was a robust, powerful example of hominism. Even if *H. sapiens* and *H. neanderthalensis* did not interbreed (which is the subject of lively debate), his diet still interests me. If my ideal, biologically ordained diet can be determined from studying other species, then it must be admitted that Neanderthal's diet is relevant. He's much closer to me than the chimp or gorilla. So what did he eat? When scientists drilled into his bones and checked his isotopes, they concluded that Cousin Neanderthal adhered to a diet of nearly pure, unadulterated, low-fiber, vitamin-challenged meat. *Growl!* I can't believe that the all-meat diet is quite right for me, either. Perhaps if I were roaming my territory all day, eating animals lean from roaming their territories all day, I could handle it. But for a sedentary human, eating modern meat, the Neanderthal path would likely end in the cardiac ward.

THE INUIT DIET VERSUS THE !KUNG DIET

Still, many swear by these ancestral approaches. Noisy bands of modern *H. sapiens* snarl over the merits of the Neanderthal steak versus the gorilla salad, and every conceivable bill of fare in between. Pulling toward the Neanderthal camp is Loren Cordain, a health scientist at Colorado State University. Cordain has made a splash with his "Paleo Diet," an eating plan that advocates lean meat, seafood, and veggies, at the expense of carbohydrates. This is an Atkins plan minus the fat and plus fruit, if you will. On the other end of the rope is a Virginia physician named Dr. Milton R. Mills, who's pulling toward the vegetarian gorillas. Mills's "Comparative Anatomy of Eating" is plastered across the Internet, looking for all the world like science as it declares the human body to be built for veggies and veggies alone. Frankly, most diet gurus achieve a shrillness that makes a primate want to cover her ears. And besides, why argue about Neanderthal versus gorilla, if living, breathing *Homo sapiens* are still out there hunting and gathering, living on wild foods? Wouldn't it be more illuminating to investigate what they're eating? I turn to studies of what modern-day foragers—remnant bands of hunter-gatherer humans—eat.

And here's what they eat: Just like we hypercivilized *Homo*s with the chopsticks and the dessert forks, no two groups of modern foragers eat the same diet. Good studies of forager diets are as scarce as hen's teeth, but it does start to look as though the human animal can thrive on any number of plans. Tanzanian Hadza humans are reported to get "the bulk" of their diet from veggies and a minority from meat. (Bear in mind that the word "meat" here means animal protein. This could be fish, frog, sparrow, lizard, chipmunk, witchetty grub, cicada, caterpillar, dog, or even something as hideous to behold as a lobster. Humans really will eat anything.) By contrast, up until quite recently, Inuit humans have obtained about half their calories from meat—and the other half from fat. A few berries, a bit of seaweed, and the stomach contents of lichen-eating caribou constituted the vegetables. That's a rather Neanderthalesque choice. Then there are the !Kung from southern Africa, who collect two-thirds of their calories in plant form and one-third as meat, which is not far from the ratio my own culture embraces.

But even if I arbitrarily chose one of these meat-to-vegetable ratios, I would quickly find the guidance irrelevant to my modern life. The quality of meats and veggies found in a !Kung territory and those in my supermarket are incomparable. Wild meat contains an average of 4 percent fat. Lazy beef cows can hit 36 percent. And the veggies I eat are domesticated species whose nutrients faded in the process of their taming.

Further complicating the comparison, if I choose to model my diet on a forager's then I should tailor my own meat eating to that of a female forager, not to the general population. It seems that the hunting males in many cultures strip out the most nourishing parts of an animal—organs, fat, and marrow—before they haul it back to the rest of the group. Even then, it appears that butchering is often conducted according to rituals and taboos that keep the same few humans well fed. Those few humans are almost never female. In 1978 a research physician summed up observations of meat sharing in one Australian Aboriginal community thus: Those fed first are old males, followed by hunting males, children, dogs, and last, females. An unpublished study of fossil *H. sapiens* from a site in Turkey suggests this is an ancient pattern of food distribution: Females there had lower levels of meat isotopes in their tooth enamel than did males. At first glance, it looks like a case of male self-preservation, but it may, in fact, say something about the female's ideal diet. Modern females on a high-protein plan like the Atkins Diet can have difficulty getting pregnant. During pregnancy, too much protein may also drive down the birth weight of a baby. Furthermore, animal liver, with its high vitamin A content, can cause birth defects or spontaneous abortion. In fact, meat is a common subject of the food aversions that human females suddenly experience in pregnancy. So it could be that foragers' meat rules serve a biological purpose: to keep females fertile. (The meaty diet of traditional Inuit women has not, however, resulted in the group's demise. So this may not be a universal problem.)

The sense I get is that the human animal can and will thrive on a bewilderment of diet plans, from all vegetable, to a near totality of meat and fat, and every combination in between. This is the standard omnivore pattern, designed for maximum flexibility. We are omnivores, through and through. Where the human diet diverges from all others, though, is in the cooking. We are cooking omnivores, and that has contributed to our success.

STARVATION, AND MY PLAN FOR AVOIDING IT

If the jaws of our ancestors can only mumble that I'm equipped to eat anything, then I'd at least like to know how much—of whatever—I should consume today. The simple answer, common to all animals, is "at least enough to live until tomorrow." To add a layer of sophistication, the answer might go on, "Ideally, enough to store some fat to get me through a shortage. But not so much that the weight of the stored energy interferes with my other daily tasks (breeding, gathering resources, running from bears)."

From the look of things in the mirror, I've struck this balance. If anything, I'm a little overprepared for a famine. But I can hardly blame my body. It's operating on the old assumption that food behaves unpredictably. That assumption was sound for millions of years. So my body is sticking with it, even though food has become more reliable during the past century.

For many other species, starvation is the predator that never sleeps. In Sweden, one study found that a staggering one in five roe deer dies of hunger. On Isle Royale, a national park in Lake Superior, wolves perish of starvation and violence (which biologists link to food shortages) at a rate ranging from 18 percent to 57 percent in a year. The snowy owls of Alberta, Canada, do a little better, with starvation killing just 14 percent of the dead owls collected in the province. But among Wyoming's female moose, malnutrition accounts for 60 percent of deaths. Starvation happens. It qualifies, in the view of wildlife biologists, as "death by natural causes."

My species has not outwitted this hazard. Though our starvation rate doesn't approach that of the Wyoming moose, neither is it zero. One out of eight humans alive today is getting less food than she needs. Millions starve to death every year. The risk for the human animal, as with many others, is highest during childhood. Among human young under five years old, about 1 percent die of starvation-related ills. That's six million kids a year, globally.

The causes of starvation in humans are identical to those in other animals, although we don't consider them "natural." One is meteorological. One year, the rains simply don't arrive on schedule, and the humans who depend on grain suddenly find themselves in the same boat as the rabbits who depend on grass and the herons who depend on tadpoles. Territorial conflict is another cause: In the Sudan, the Arab Janjaweed push the Black Africans off their traditional territory, and the Black Africans promptly begin starving. In Yellowstone National Park, the same fate awaits a weak wolf that a stronger pack bullies away from the elk herd. And both of these catalysts, meteorology and conflict, burn hotter when they spark in a dense population. In Ethiopia, the human reproduction rate has been growing since 1960. Each generation of females is producing more offspring than the generation before. The population is rising faster than almost anywhere else on Earth, and it's also bumping up against some of the world's most erratic weather. Ethiopia's droughts have become legendary, and as more and more *Homo sapiens* crowd the unstable habitat, the starvation risk soars.

In addition to outright death, the human species carries a heavy burden of malnourishment from insufficient food. An estimated quarter million to half million human young go blind every year from lack of vitamin A. Another twenty million mentally retarded *Homo sapiens* walk the earth today because their mothers' diet lacked iodine. Half the world's pregnant females are anemic from inadequate iron, which raises their risk of dying in childbirth. All things considered, my species isn't quite as good at chasing down sufficient nutrients as it appears to be here in the chubby developed West. Starvation still preys on us. And as we'll soon see, even Western chubbiness is no guarantee of sufficient nutrition.

But first, let's look at my natural defenses against starving. There's a good reason that I yearn for fettuccine Alfredo and chocolate. Every cell in my body is in a near-constant state of hollering for high-calorie food. My body wants to be bigger than it is today. Therefore my cells lobby for more sugar, more fat, more food. A pathetic minority of cells barricaded in a corner of my brain is all that stands between me and a plate of nachos right now. They're the cells that deal with reason, cause and effect, and long-term planning. They, too, slave day and night, to stop me from diving on the butter.

Why fat and sugar? Why don't I crave salad? My body is lagging behind the times. For the first few million years of hominid existence, salad was everywhere. You had to kick it out of the way just to get around. By contrast, energy-rich foods were either too seasonal or too fleet-footed for convenience. Most nuts, rich in oil and protein, were only on the menu for a few weeks a year, and competition was spirited from weevils, rodents, bears, and birds. Ripe fruits, too, with their jolt of high-energy fructose and glucose, were sporadic and fought over. Wild meat was available year-round, but it was lower in calories than nuts, and more evasive. The situation today is not so different: As I look out my office window, I see tons of edible greens—sheep sorrel and chickweed in the lawn, daylilies and tulip bulbs in the gardens; even the oak leaves will yield a smidgen of sustenance to a human in need. But the fruits—chokecherries, crab apples, honeysuckle berries, sumac berries—will ripen suddenly, and just as suddenly be devoured by birds and insects. As for nuts, many of the acorns will fall already punctured by weevils. Squirrels and jays will rush away with the survivors. Meat is more plentiful than nuts in my yard—squirrels and pigeons look most promising—but even these dopey, half-domesticated animals are so spry I suspect salad would constitute 99 percent of my diet if I were relying on the fat of the land and a few handmade tools. And, taking spinach as a stand-in for wild greens, I'd have to eat a pound of salad to capture the calories in a small chicken breast. I'd have to chow five pounds of slaw to get through the day. Even then, mere calories are only half the battle. I need protein, carbohydrates, and micronutrients, too.

So it's no surprise my body has an instinctual attraction to foods that will speed me toward my daily quota of calories. This feature is not unique to humans. Honeybees devote their time to those flowers with the sweetest nectar. A coyote who kills a fawn will strip out the fatty heart, liver, and other organs before carving into the lean meat. When a polar bear bags a seal he'll often eat only the skin and blubber, leaving the muscle for scavengers. When prehistoric Plains Indians drove buffalo over a cliff in greater numbers than they could process, they, too, focused on the rich tongues and neck humps. Animals tend to be efficient in the execution of all their duties, because to noodle around on the job exposes them to danger. Accordingly, when my senses register the proximity of a Chunky bar, my strongest urge is to snatch it up and get it down the hatch before, A: it gets moldy; B: it's eaten by bears, C: I'm eaten by bears.

My salt cravings have a similar history. My body needs somewhere between 115 and 500 milligrams of sodium in a day. To get 200 milligrams from a vegetable diet, toss (in a bucket): an entire head of lettuce, a cup each of cabbage, green pepper, asparagus, nuts, cooked kale, mushrooms, turnip greens, and Brussels sprouts, plus an avocado and a baked potato, and garnish with a pint of strawberries and cherries. And happy chewing to you. If you live in a hot climate or are prone to moving about vigorously, then eat at least two salad buckets a day to keep pace with your sweat loss.

So, once again, it's meat to the rescue. Or animal blood, actually, which runs about 1 percent sodium chloride. Fish, especially ocean species, also deliver salt, as do seaweeds and other marine munchies. I crave them all. This craving, like the fat craving, is widely shared. Although salt is a common mineral in the earth's crust, deposits are widely scattered. Where they do break the surface, animals have usually beaten a path to them ages since. In western Kenya, grazing animals including elephants make a regular trek to Mount Elgon, where a deep, dark cave beckons them with the salt in its walls. In darkness, braving hyenas and leopards, the herd animals gouge out salt, then run for the safety of the savanna again.

One reason cravings are so strong, and that it's so joyful to yield to them, is that they tap into the same brain chemistry that will get a human hooked on cocaine or alcohol. When the sugar from a Chunky bar hits my bloodstream, opioids of my own making flood my brain with chemical happiness. I've eaten enough Chunky bars by now to get my brain addicted to these opioids. I need no heroin, only another Chunky. My brain will produce its own high. But deprive it of Chunkys, and it will start to produce a craving, a gritty, chemical need. And at least in rats, who stood in for humans in a recent study of addiction, that same opioid surge encourages animals to seek more fat, salt, alcohol, and some drugs. Human brains, whether they're plied with alcohol or antidepressants, act much like rat brains. In fact, all vertebrates (animals with backbones), and especially mammals, share this opioid system that sparkles in response to calories. Scientists presume that's because we vertebrates also share a need for high-energy foods and rare vitamins and minerals, and the reward system evolved very early in animal history to make sure we pursue them. We're all in this together, a twitching bunch of food junkies jonesing for the next batch of brownies.

My cravings for sugar and fat are steely jawed, and deservedly so. For most of hominid history, starvation was nigh and calories elusive.

The only means we had for preserving food was to eat it and convert it to fat. Those humans whose brain chemicals whipped them most doggedly toward high-cal meals were rewarded by reproducing regularly. A famous little carving called *Venus of Willendorf* suggests that at least thirty thousand or forty thousand years ago, *Homo*s recognized the value of a physique swollen from the diligent pursuit of lunch. We'll probably never know what the artist had in mind when she or he shaped the limestone figure, whose breasts spread over a well-stuffed stomach and whose buttocks bulge with stored energy. But given that many Stone Age icons of the human female seem to celebrate an ability to subsidize babies, it's reasonable to speculate that Obese Venus and her many graven sisters were symbols of fertility. Reproducing is a primal drive, and it's fueled by fat.

This limitation remains with us. When the human female's fat shrinks a few points below 20 percent of her weight, the ovaries simply decline to sacrifice an egg. Why send one down if times are so hard that the host can't even feed herself, let alone a fetus? Nature abhors waste, and so ovulation is suspended until there is more energy in reserve. In wealthy countries, starvation-induced infertility strikes marathon runners, gymnasts, and ballet dancers. And in less wealthy cultures, starvation infertility is quite common among the sixty or seventy million females in the world who are unintentionally undernourished. Naturally, it is also a fact of life for other animals. Growing a baby animal demands many extra calories, and a body won't initiate the process if the odds of success are skinny.

My unwavering quest for fat and sugar is a wonder of evolution and a souvenir of all mammals' ability to end-run adversity. It is my creature heritage that allows me to gulp that fettuccine Alfredo, then swallow a block of chocolate pastry, and tip a stream of cream into my coffee. It is animal nature that allows diminutive speed eater Takeru Kobayashi to cram fifteen thousand calories' worth of hot dogs into his digestive tract in twelve minutes. And impressive though Takeru and I may be, we're nature's featherweights in the calorie-slam event. Among our fellow creatures, a big Siberian tiger can eat 15 percent of his weight in a sitting. I'd have to eat twenty-two packages of hot dogs to keep pace. The dainty king snake has been known to attempt ingestion of a corn snake four times its size. And the noble polar bear can eat 20 percent of its weight in seal blubber in one meal. For me, that would be about thirty pounds of butter (93,600 calories' worth). I bow to the superior wolfing of my peers, and to their efficient foraging techniques. But here's a warning: We're gaining on them.

OVER-OPTIMAL FORAGING: THE FEEDING FRENZY

Optimal foraging is the state of feeding toward which all animals strive. Feeding always has "costs," as biologists say. It burns energy, and it may expose an animal to thieves and predators and lightning bolts. To achieve optimal foraging is to maximize the nutrient intake and minimize the costs. For the leopard, that means hunting at night when the gazelles can't see you and the lions are too sleepy to beat you up. For the squirrel, it's working in a forest with ten acorns per square foot, rather than a forest with one acorn per square foot. For a human, it means burrowing deep into your sofa while persuading the mate to bring a bowl of ice cream. With all due modesty, we *Homo sapiens* have planted our big, fat flag on the bulging summit of Mount Optimal Forage.

Consider the time other animals spend procuring their daily bread. The orangutan eats for five hours a day. The manatee spends six to eight hours. Elephants grind their grub sixteen hours a day. Dust mites eat skin around the clock, without cease. But I, due to the ability of *Homo sapiens* to preprocess food, am able to devote as little as fifteen minutes a day to opening packages and giving their contents a cursory chew. Add a weekly trip to the supermarket for more packages, and I'd still spend less time foraging in an entire week than an orangutan spends in a day.

This doesn't represent the primeval human state. My space-age foraging strategy began with the invention of herding and agriculture, about ten thousand years ago. Agriculturists, instead of chasing plants and animals across the face of the earth, raised their protein and carbohydrates in the dooryard. Convenience food was born. The domesticated fruits and vegetables grew both in size and sugar content. Cows, goats, and other animals that we bred for food lost their suspicious natures and gained body fat. Gradually the diet of agricultural humans grew richer in starch. Then about a thousand years ago the traditional sweetener, honey, was joined by a new one made from the juice of the giant grass sugarcane. This was energy in one of its purest forms. Speeding forth to the late 1800s, the invention of the "roller mill" made white flour affordable to millions of humans. The ensuing white bread saved *Homo sapiens* a great deal of chewing. Soon thereafter, scientists presented still another high-energy food—corn syrup—to the public, for even less money than sugar. Then in the 1960s, chemists found a way to make corn syrup three times sweeter, and—*Stop the roller mill! What are we eating here?*

The omnivorous ways of *Homo sapiens*, and our built-in drive to eat fat and sugar, are normal animal features. There's nothing weird about them. But lately our talent with tools has made the process of foraging so efficient that we can consume quantities of food that are fatal. Our antique bodies have yet to catch up, yet to develop a means of saying, "That's enough." The result is that, for the first time in a century, the life span in many human groups may be shrinking. Human foraging has developed a suicidal aftertaste.

This is freakish biology. Other animals don't eat themselves into the grave, unless they've been domesticated, captured, gene modified, or otherwise forced into an unnatural situation. Some animals will put on weight for a specific purpose. A bird facing migration might stow a few ounces of fat but will burn them off in flight and return to normal weight. A woodchuck preparing for hibernation will pad herself with sufficient blubber to survive the winter but will be slim by spring. Humans are different. We can easily double our body weight, then maintain that fat even as it destroys our health.

There were early hints that our optimized foraging would backfire. When scientists look for cavities in the teeth of fossil humans from before the dawn of agriculture, they find nearly none. But the taming of grass seeds led to a diet richer in mush. Mush sticks between teeth, where it instigates rot. Malnutrition also sets in, launching a trend that continues today. When scientists measure the skeletons of *Homo sapiens* from before and after the advent of agriculture, they find the bones of early farmers stunted and stressed. They find that offspring died in greater numbers. But they were born in greater numbers, too. And despite the rough start, farming took hold worldwide. It is rare for a human group to adopt agriculture, then forsake it to return to the forager's lifestyle. It may happen when invaders push a subsistence group off their cleared land and back into the forest, as occurred throughout the Americas a few centuries ago. But the vast majority of humans now base their diets on farmed foods.

The trajectory toward effortless feeding has shot up in the past century. The "hands-on" work of foraging is now conducted on huge centralized farms, and processed food travels cheaply across thousands of miles. Every year, millions more humans migrate from a rural territory to an urban one, where the only place to find food is a market. Confronted with the modern calorie, whose biological costs are negligible, the human body efficiently captures it and moves on to the next calorie.

Obesity is the opposite of starvation. It is optimal foraging taken past the optimal point. It is evolution caught unready to cope with the efficiency of its own handiwork. Human feeding has become fatally efficient.

In obesity, a second feature of the human antistarvation system comes into play. First, my cravings drive me to eat high-calorie foods. Then, if my body detects starvation's approach (when I start dieting, for instance), my metabolism rises to meet the challenge—or, more accurately, falls to meet the challenge. All my cells slow down, practicing fuel efficiency body-wide. My fat lasts longer, seeing me through the "famine." It's ingenious! When there's food, I gobble it like there's no tomorrow. When there's none, I turn back the thermostat to make my stored fuel last for tomorrow and tomorrow and tomorrow. This makes the intentional shedding of fat extremely difficult, of course. Not only must I do battle with the cravings, which wail from every cell; I must also combat my body's desire to embrace torpor. Once, it was a marvel of adaptation. Now it's a miserable maladaptation.

Homo sapiens across the globe are giving up the battle against their outmoded bodies and are growing enormous. Among the hominids in my culture group, three out of ten adults are obese; another three are overweight. Many other cultures are bulking up, too, as over-optimized foraging spreads around the world. Some cultures have shown an astonishing aptitude for storing fat. Samoa and Nauru, Pacific Island cultures, have achieved obesity rates that range from 60 percent to 80 percent of adults. Although genetics, and perhaps diseases, play a role, cultural pressure also influences the obesity epidemic. In parts of Africa, for instance, girls and young women are force-fed to attain a state of fatness that will render them irresistible brides and bounteous mothers. This tradition is on the wane.

Even human young are over-harvesting energy. As parents cater to the cravings of their offspring, the young are ballooning in record numbers. The diabetes, heart disease, and other risks associated with youthful obesity are so severe that some researchers now predict a shortened human life span, both in my culture and in others where the young are becoming obese.

Perhaps the most unexpected trend *within* this freakish trend is the appearance of humans who are both obese *and* starving. A great flurry of research on obese young has found they're often deficient in vitamins A and E. Whether this reflects a diet too low in plants, or a chemical reaction within fat tissue, or a combination, isn't yet known. Another study has correlated the soft-drink consumption of young humans with both rapid weight gain and deficiencies in phosphorus, protein, magnesium, calcium, and vitamin A. And yet another study reports that iron levels are low in obese children; the authors speculate that kids may be replacing nutritious foods with caloric ones.

One reason for this paradox is the old salad problem: The human body has not evolved to crave salad because salad has never been so rare that we needed an opioid motivator to seek it out. We evolved to crave the rare things, the sweet and oily and salty things, which quickly fill the tank and even buy some time for tomorrow. We have evolved to overeat when the opportunity presents itself. Today that opportunity is ever present.

Another reason that the human can be simultaneously overfed and undernourished is that starvation does still prowl the planet, most often following on the heels of poverty. Today, a calorie of sugar or fat is much cheaper than a calorie of protein or fresh vegetable. So in areas where humans have minimal resources to trade for food, they often choose to buy the largest number of calories for the smallest amount of money. It's optimal foraging, in a sense. But because nutrient-dense foods like spinach and fish are too expensive to satisfy the optimal-foraging formula, those foods are avoided.

Obese *Homo*s now outnumber underweight ones worldwide. Our rapid transformation from a "normal" animal to a fat one is unparalleled. There is no analog in nature. No other animal will eat to the point of harming itself. Even among hunter-gatherer humans, obesity simply doesn't happen. In fact, the only animals that will eat themselves silly are the ones we've domesticated. By protecting and feeding them, we've removed the costs that would naturally accompany their feeding. My own dog, for instance, would cheerfully eat lard until his pancreas exploded. But not so the crows in my yard, the mice, the squirrels, or even the hibernating woodchuck. Even that little fatty will store only enough padding to see her through the winter. To spend more time exposed to hawks and dogs would be biological nonsense.

■ ■ ■

So at last: What is the diet of the human animal?

All rhetoric aside, these are a few of the things that the human animal eats: acorn, ant, artichoke, barley, bean, beetle, cicada, coffee, cow, date, daylily, dolphin, edamame, eel, egg, fennel, fern, fungus, garlic, goat, Gouda, haddock, hare, horseradish, ikura, Irish moss, iroko, jackfruit, jaguar, jicama, kamut, kangaroo, kiwi, lark, lemon, lentil, mango, mouse, mustard, nasturtium, nectarine, needlefish, onion, opossum, owl, pansy, papaya, paprika, quahog, quince, quinoa, rat, rhinoceros, Roquefort, shark, sheep, sumac, tamarind, tequila, thistle, ugli fruit, uinta chipmunk, umi, vanilla, Velveeta, vicuña, walnut, whale, wine, yak, yeast, yucca, za'atar, zebu, and zucchini. If this does not describe an omnivore, I don't know what does.

Happily for those humans more idealistic about their foraging, our species can squeak by on subsets of food that might include mainly the chipmunks, sheep, and dolphins; or mainly the yucca, jicama, and barley. And happily for those who hunger more for exploring the world than for dietary idealism, the omnivorous human can go anywhere. We'll always find *something* to eat.

Two points separate me from the other omnivores, however. First, although my species will scrounge with the best of the bears and skunks, most of what we collect we'll want hot. Most of the foods on the list above are things I prefer cooked.

And also unlike my fellow creatures, I'll eat more than is good for me. I am genetically directed to overeat when I can. And while other animals might shelter the same genes that steer them toward high-benefit foods, they don't give those genes the run of the house. They have other priorities, like avoiding being eaten themselves. I, like a domestic dog or a chimp in the zoo, eat without fear of predators. It is only fear of social condemnation that keeps me from heading out this minute to forage in the Chunky fields.

And that social condemnation, so painful for a social animal to contemplate, may itself be losing value. If everyone is fat, where's the social risk in overforaging? Once again, our unusual biology is leading us off the path trodden by all other creatures and into terra incognita.

7

LOOSE AS A GOOSE:

REPRODUCTION

The life cycle of *Homo sapiens* is plodding. The animal doesn't achieve breeding age until late in its second decade, although with improved nutrition that age seems to be lowering. When the female does conceive, her reproductive capacity is suppressed for up to five years: nine months for gestation, and the balance for nursing an offspring that is incapable of gathering food, transporting itself, or indeed producing any useful behavior at all. This burdensome investment is typical of all the great apes. Mercifully, litter size is usually limited to one.

Uncharacteristically for a mammal, breeding humans form a pair-bond which, depending on the culture, might last through the early years of one offspring or for life. Mating decisions are based on a vast number of considerations, including behavior, age, race, and perhaps physical symmetry and body odor, in addition to myriad cultural factors. It is not uncommon for both sexes to pursue a dual mating strategy, forming a pair-bond with one partner, but seeking mating opportunities with additional partners as well. Polygamy is also common. It is not entirely clear why the human has been described as monogamous.

Birth in this species is traumatic for mother and infant, due to the infant's enormous skull. In some cultures the rate of maternal death is one in twenty. This underscores how heavily the risk of reproducing rests on the female of this species and helps to explain why she approaches mating decisions with a studiousness that can be difficult to detect in the male. Lifetime fecundity varies with culture, averaging two or three offspring per female among urban humans, four or five among subsistence humans, and eight (which is probably the natural maximum) among farming cultures. Modern tools (infant formula, test tubes) facilitate greater fecundity, for those who seek it. A separate tool kit permits humans to limit their fertility, or forgo reproduction altogether. Although the latter choice would be anathema to any other creature on Earth, this is increasingly common behavior.

The human father participates in the care of offspring produced within the pair-bond, but any extra-pair offspring he might sire are typically raised by the female in question and possibly her mate—this male being known in the parlance as a "cuckold." The term derives from the word "cuckoo," a bird so accomplished at slipping eggs into others' nests that she need never construct a nest of her own.

THE REPRODUCTIVE WINDOW OPENS

I first experienced the desire to mate when I was eleven. My neighbor "Dick" was forming a puppy pair-bond with the diminutive "Jane," which cast a new light on him. He was more than the roughneck down the road, I discovered. He was a male. And I wanted him to be my male. Never mind that my body would represent an extremely poor investment of male sperm for another three or four years. Never mind. Like a rocket that takes all day to launch, I had begun to pressurize the tanks and prime the pumps in preparation for what was sure to be a dazzling career.

Like my mother before me, I was slow to reach breeding age, stretching taller as the females around me hit puberty and ceased growing. I trudged along, frustrated by the fact that the rising estrogen seemed to be pressurizing my brain much faster than it could inflate the rest of me. With knobby knees still jabbing holes in my jeans, I studied my babysitter Marguerite's curves. I wanted the boy down the road, and I wanted to be Marguerite. It would be another three or four years before my reproductive system was completely online.

The human female reaches puberty between the ages of nine and sixteen. (The age depends mainly on heredity and can be accelerated by good health and nutrition.) But even after that, most humans benefit from a trial period when, despite the appearance that the plumbing is cycling, they are nearly infertile. I probably menstruated for a year and a half before I started launching eggs down the fallopian tubes toward the uterus. And still I wasn't ready to breed. Sure, I could have conceived any time I ovulated. But until the human female is eighteen or twenty, her body isn't really ready to carry a fetus. Had I conceived, I would have risked dangerously high blood pressure and offspring that were perilously early and small. And, of course, I might have discovered too late that my pelvis had yet to grow wide enough to permit the passage of an infant's skull.

The average male hits puberty two years later than the female. My American classmates began to produce viable sperm at thirteen or fourteen (the end of junior high school, or the beginning of high school). With their soft cheeks and narrow shoulders, they could not be mistaken for adults. And all boasting to the contrary, I don't imagine they scored many cops. (Biologists coined this abbreviation for "copulations." The contrast of the tidy word with the awkward deed it describes pleases me.) In fact, males under twenty can claim responsibility for just 3 percent of all American pregnancies, according to a 1990 study. Like my young female friends, my male cohorts were in a gawky and somewhat protected state of sub-maturity.

Young humans often take advantage of this phase to practice, as it were, the task for which biology has been preparing us since the day we ourselves were conceived. We begin to play at reproducing. Alas for me, my biology was still out of sync with my psyche. Though I would have been delighted to pair-bond at the age of fifteen, the males around me were having none of it. For starters, I was tall at an age when the males were still catching up with the females. On top of that I was a dominant, aggressive female. My pair-bonding career got a slow start.

Speed through three hundred more ovulations, and here I sit in my early forties with a uterus that is increasingly hostile to embryos. My opportunity to build a lasting biological legacy is coming to a close. Technically, I could still reproduce. But it would take a great many attempts to conceive, and then my uterus might reject two or three embryos in a row. If I could bring a fetus to term, its DNA might be so degraded that it would represent a biological dead end after all, unable to do its own reproducing.

Had I been a dutiful organism, with the aid of feeding formula I might be overseeing 15 offspring by now, each shuttling my genes toward the future. This sounds like quite a legacy, but it's pocket change for many mammals. Had I been a dutiful rat, I might be the mother of hundreds. A truly exemplary rat can breed once a month for three years, producing up to 22 pups at a pop, for a total of 792 progeny. Even an average rat can bring 20 pups to adulthood in a two-year life span. But rats, like rabbits, lemmings, mice, and squirrels, are in a category of animals I think of as grocery species. They're the staple foods for a host of predators. To move their DNA down the line, these grocery animals have to produce a zillion offspring in hopes that the eagles, wolves, owls, coyotes, foxes, fishers, weasels, stoats, and skunks will be sated before they catch the zillionth baby.

I am too large to be a grocery species. Like most large animals, my species breeds occasionally and slowly. A human female typically raises between two and eight offspring in her fertile lifetime. Urban females, for whom offspring are expensive, are usually at the low end. Farming females, whose offspring can contribute to their own support at a young age, are at the high end. Subsistence cultures, in which females must carry infants while foraging, average four to six. The human range of fecundity is in line with that of the chimpanzee, the bottlenose dolphin, and my local black bear.

I, personally, failed to produce so much as a single pup. Now here I am at an age when my most ambitious peers are grandmothers. I'm looking at another four or five decades of life with no progeny to brighten my days. There are so many things wrong with this. For one thing, it throws a wrench into the theoretical works. My biological raison d'être was to reproduce. Without that, what is the meaning of my biological life? Almost nil. It's not *completely* nil, because I could still advance my genetic cause indirectly, by assisting the offspring of my siblings. Because my sibs carry some of the same DNA I have, they're a second-best investment of my resources. I could also invest in the offspring of my cousins, although I share even less DNA with those relatives. However, in my generation of fifteen cousins, only half of us ever bred. Neither of my siblings is among them. So yes, barring my sudden, crucial investment in a cousin's offspring, the meaning of my own, personal, biological life is, actually, nil.

Another problem with my fruitlessness is that it messes up my narrative line. If I had bred successfully, I could now cast a line out to my old age, when I would be caring for grand-offspring, long after my own reproductive window closed. Gazing at the horizon, I could comment naturally on the remarkable length of a human life span. Now I have to force it: Mine is one of a tiny number of species in which the females live long, looooong past their reproductive years. Virtually every other animal—badger, beetle, boa constrictor—reaches the end of her breeding life and her breathing life simultaneously. Breeding and living are synonymous. When one ceases, so does the other. But humans live on, with females moving into a strange, infertile phase called menopause. Many theorists have ventured to guess why. None has yet proven her case.

The olden, golden theory is that grandmothers are a new phenomenon, that humans who live without refrigerators or tetanus shots only make it to about forty. It's only in the past century that the average human life expectancy has moved much beyond that number. Subsistence cultures like the Amazonian Yanomami still have short life expectancies. Therefore, old females are a by-product of the modern age. But that theory suffers from number crunching. It's the high rate of infant mortality that drags the average life span down. All those infants whose life spans are a single year make it appear that everyone lives a shorter life. So let's crunch the numbers differently: Of those females who *do* survive through their breeding years (roughly age forty), how much more life can they expect? And that average is about another twenty years. That (along with quite a large number of grandmothers alive and well in subsistence cultures) does no favors for the olden theory.

But we have other proposals to choose from. Also in contention is the notion that menopause is a by-product of the human animal evolving such a long life span. The body runs on so long, in other words, that the breeding gear just wears out. I don't like it. Plenty of long-living animals—the elephants (seventy years), humpback whales (fifty years), and the African gray parrot (fifty years), for example—keep breeding nearly until they drop. Something makes us different from them.

The new crowd-pleaser is the grandmother hypothesis. This posits that a female who has already set her genes in motion down the river of time can give the vessel one last push if she sticks around to help out her offspring's offspring. And sure enough, if you visit the Hadza culture of Tanzania, you might see something like this: A breeding-age female, with a toddler in tow, is moseying through the desert collecting easily harvested plants. Meanwhile, Grandma is whaling away at the hard ground with a digging implement, harvesting roots that will nourish a growing body. Mother may be eating for two, but it's Grandma who's foraging for two. Among the Hadza, the offspring whose grandmothers spend the most time gathering food weigh the most. Anthropologists are collecting similar scenes from other subsistence cultures, too.

I'm inspired to conceive my own theory, because I'm not crazy about any of the others. The grandmother hypothesis is okay, but even among hunter-gatherers, tons and tons of breeding females raise a normal brood without the aid of a mother or mother-in-law. What I *do* like is the idea that human offspring take more than ten years to grow up. I've encountered legitimate and degree-holding theorists who argue that offspring need parental guidance and assistance for only their first ten years of life. Having been eleven years old myself, I respectfully disagree. To thrive as a hunter-gatherer, which all humans had to do until fairly recently, required a knowledge of plants, animals, weather, shelter, tools, and human nature that I just don't believe the average *Homo* can absorb in her first ten years. Sure, millions of young children work hard, in both subsistence and developing cultures around the world. But working is easy. The trick is translating labor into the necessities of a life. I was cooking for my family at the age of ten, but I would have been hard-pressed to assemble all the shelter and tools necessary to do the job, even if I could round up edible raw materials. And as for defending the territory from invaders and raising my own offspring: no way. I reckon I needed the protection and instruction of adults until well into my teens. My culture agrees, declaring that all humans under eighteen shall be supervised. And throughout our species, the average age of marriage hovers around twenty, signifying that younger humans aren't ready to run a household just because they can reproduce. So there's my theory: Humans, and females in particular, evolved to live long enough to guide their final offspring through their teen years.

(And grandfathers, you ask? Research has ignored old men until very recently because males were thought to remain fertile throughout their life spans. We now know the male's fertility fades, too, and that his aging sperm raise his offspring's risk for schizophrenia, autism, and other psychiatric ills. The barest hint of research suggests that old males, too, might contribute to the survival of grand-offspring, delivering a parting push to their own genetic rafts.)

I mentioned the possibility of other species reaching grandmotherhood. There are a few. A certain guppy lingers on after her last brood, and experiments show this is strictly a side effect of evolving an unusually long life (for a fish). The pilot whale spends one-third or more of her life in postmenopause, and does, in fact, help with the grandkids. And recently, a gorilla named Alpha made the news for entering menopause at the Brookfield, Illinois, zoo. But Alpha is just a good advertisement for modern medicine: Captive animals usually live much longer than their free-running kin, who rarely brush their teeth and never floss.

Before we close this breeding window, I need to revisit a curiosity of human reproduction I first mentioned in chapter 3. I cannot count the times I was living or working with another female and discovered that our fertility cycles had synchronized. It doesn't always happen when females congregate for months at a time, but it is a common experience. Scientists don't know why. Two competing theories illustrate the split personality of our species. There's a cooperative proposal and a competitive proposal. The first camp argues that when females breed in concert, they end up with infants of the same age. This makes for a group of mothers who can trade babysitting and wisdom, and their offspring can learn social rules from one another as they grow. Bottlenose dolphins, for one, demonstrate that mothers who form this type of cooperative nursery raise more offspring than do mothers who go it alone. But the human is fertile every four weeks, and the offspring very slow to grow. I find it hard to believe evolution would punish those whose infants were just two or three weeks younger or older than their peers'.

The more cynical view is that human competitiveness drives females to synchronize their fertility. If one female in the group should attract a magnificent male to the neighborhood during her fertile phase, the others may want to be ready to exploit the gent's fine genes. Mice appear to take this approach. All the females in a given group quickly become fertile at the smell of a male. But elephants seem to have it both ways. The females in a herd synchronize their fertile period. Then when the time is right, they collectively hoot and holler to summon distant males. But the dominant matriarch may monopolize the prime male, leaving the others to sort through her castoffs. Research indicates that the matriarch tends to give birth first, in the lush days of the year; subordinate females drop their calves during drier times and lose more of them. There's nothing cooperative about that. I suppose both could be working at once in the human animal, too: Theoretically, it would be in my best interest to have the only, and best-fed, offspring, but if I can't monopolize the copulation opportunities, at least I can get some help with the diapers and feedings.

I can identify both pros and cons in the timing of my own birth of a few decades ago. My mother began breeding shortly after her older sister; their brother was right behind them. The hand-me-down clothing I grew up in was a huge benefit to my mother, as was the babysitting the three mothers exchanged. But there were also times when all the mothers wanted babysitting from my grandmother, and my gracious Gramsie could only handle so many of her ten screamers at a time.

Ah, my Gramsie. And her Gramsie. And a thousand Gramsies before her. It's unsettling for me to be the very last twig on a tall tree of humans who for thousands of generations, since the dawn of life, actually, have diligently bred, each optimistic that his or her own DNA would reach eternally forward. And then there's me. The dead-end twig of our tree. Sorry. I really am. From the first single-celled organisms to me, it was a glorious run. But this is more than an aesthetic failure. It also exemplifies the bizarre state at which my culture has arrived: The human animal competes in the territory winning and calorie chasing as though it planned to produce a thousand progeny. Then, in my culture, many of them produce none.

This makes a lot of human behavior look silly. But more on that as the silliness crops up.

THE PROCESS OF PAIR-BONDING

My first serious attempt at forming a pair-bond, let's see. This is awkward. Let's call him Homoboy. He could, after all, have been anybody—or at least anybody who had the correct body odor, facial bone structure, and degree of bodily symmetry. He must have had enough of those things, because for a while Homoboy took the shape of an obsession in my seventeen-year-old brain.

I'll defer the mortifying postmortem of this particular pair-bond by considering why it is that humans take only one primary mate at a time. The females of most mammals don't attempt to pair-bond but have evolved to perform as single mothers instead. Of the species who roam my backyard, none attempts co-parenting of the offspring. The possum is a single mother, as is the gray squirrel. The skunk takes the single route, and so does the raccoon. The deer mouse and the Norway rat, even the deer, moose, and black bears who pass through, all are animals who rendezvous with a mate for a minute or a month, then go their own way.

But for a few species, often those with spectacularly needy offspring, it takes two. It's a rare lifestyle among the mammals. Only about 5 percent of all mammals form couples—and most of those agree to a contract that lasts for only one year or one brood. So who are these elusive pair-bonders, and why do they do it? We (reputedly) monogamous creatures are unlikely bedfellows. We include a few species of antelope, a sampling of bats, some foxes, a few American monkeys, the South American giant otter, the northern beaver, various seals, and a handful of rabbitlike Central and South American rodents. And us. It's not a popular lifestyle choice.

And what's the best explanation for why any creature would attempt to tolerate another creature for an entire breeding cycle, let alone a lifetime? That's another conundrum. The best guesses include:

Needy offspring: In some species, the young require so much care, for so long, that one parent simply can't do the job alone.
Territoriality: In species where the male defends a territory, it behooves a female to pair up. That way she (and her offspring) can pillage the resources in his domain without the strife of fighting for them.
Predators: If the female and her offspring are persecuted by predators, then evolution will sometimes promote those males who stick around to help protect their offspring.
Infanticide: In a species where males kill offspring that aren't their own (mice, lions, chimpanzees, and many others), offspring will repay an attentive father by surviving to tell his genetic tale.
Mate guarding: Because uteruses are a limited resource, a male who monopolizes one can sometimes father more offspring than a male who chases every opportunity but is always in competition with other males.
Efficiency: Migratory birds who nest during the short northern summer can gain a few days for fattening the kids if they forgo the annual mating formalities. Staying hitched saves time.

The reasons postulated for the monogamy of my own species overlap with those proposed for other animals: Human young do require constant care, and for years, not weeks. Maybe I need a partner to protect my food source, so I can reliably produce breast milk for our offspring. Perhaps a male would reduce the likelihood of lions preying on our offspring, or fend off an interloping male who hoped that killing my young would cause me to mate with him. (Hey, it works with chimps.) It could be that a pair-bond will keep me faithful to one male even when he disappears for a four-day hunting trip. Or the human pair-bond may have evolved to benefit entire communities, by reducing the friction and violence between males—in the same way that clear territorial lines reduce violence among animals. And, of course, if a male guards me even after our issue issues forth, he increases his odds of fathering my next issue, too. Perhaps a mixture of these motivations inspires monogamy in my species. Then again, perhaps not.

I'm not monogamous at all, that's another possibility. When anthropologists analyze hundreds of cultures around the world—big ones like mine, along with small, rare ones like the Hadza—they declare monogamy to be the minority path. They're not talking about a head count, in which case monogamy clearly prevails. They're saying that of 1,154 human cultures described thus far, only about 100 espouse the one-mate-at-a-time plan. All the others—Islam and Mormon sects among them—condone polygyny (many females) or polyandry (many males), even if very few humans actually practice it. The point: When humans confront the question of how to mate, the overwhelming majority of cultures conclude that multiple mates are acceptable. Therefore, *Homo sapiens* may not qualify as a monogamous animal.

Monogamy certainly claims no adherents among our closest relatives, the chimps and gorillas. Chimps live in big, sloppy groups of males and females. When a female is in estrus, she'll usually mate with multiple males. Monogamy might occur if the dominant male in the troop wants to monopolize a particularly arresting female. In that case, he'll guard her for the entire ten days that she's fertile. A lesser male will sometimes try a different route to monogamy, called "consortship." He'll befriend a female and convince her to take a trip to the hinterlands of the troop's territory. If he can charm her, or bully her, into camping out during her fertile days, he may be the only contender for her uterus. Gorillas are more strictly polygynous. A silverback male monopolizes a group of females and brooks no hanky-panky with minor males or tall, dark strangers.

Well, if my species isn't monogamous, at least my culture is. My community raised me to breed with one male at a time, like a good giant otter or snow goose. And so the pair-bonding continues.

What was it that made me want to bond with my inaugural male, Homoboy? At that stage of my reproductive career, the primary information I gathered about any prospective mate was visual. Homoboy was taller than most of his peers. In general, human females prefer to mate with tall males. And tall males, in general, finish their mating career with more offspring than shorter ones. For males (but not females) it pays to be tall.

My target was also handsome. I mean this in the sense that a heavy dose of testosterone was causing his mandible to grow into a broad V of bone that came to a blunt end at his chin. Such males, with the large mandible and cheekbones that testosterone produces, are famously attractive to females. There is good reason to find them dashing. Testosterone actually handicaps a body's immune system. But high-quality males can afford compromised immunity, because they're so formidable to start with. It's akin to a millionaire shrugging off a big property-tax bill: It just doesn't hurt so much when you're rich. In biological terms, Homoboy's mandible and cheekbones were a "true signal," testifying to superior health. A stag's branching antlers are another true signal: The more complex the antlers, the more vigorous the animal's sperm. Although at seventeen I wasn't conscious of it, Homoboy's robust jaw was a truthful advertisement of his overall robustness. If I was after hearty offspring, I was barking up the right tree.

As for me, I was a commensurate target. In addition to having the estrogen-stunted little mandible considered de rigueur in the human female, I also had a popular body shape. The latest word on the male checklist goes thus: My admirer would have spent the most time analyzing my face. But he also checked my overall ratio of volume to height. Had I weighed too much for my stature, I'd have been rejected. Next, he checked the distance from my chin to my waist, which paradoxically yields an indication of leg length. And leg length is important. Legs do their growing in childhood, and short ones would have indicated that I had been poorly nourished. Poorly nourished children mature as sickly adults. To date, science has linked short legs with liver damage, heart disease, diabetes, and even dementia in the mature human. Finally, a glance at my waist-to-hip ratio would have told Homoboy whether I was storing my fat on the hip as young females do, or depositing it at the waist like a female at the end of her breeding career.

At least, that's how Homoboy *might* have proceeded. Researchers learn this stuff by showing photos or drawings of females to panels of males and asking which they would prefer to mate with. For the most part, these experiments have been conducted in Western or Western-influenced cultures. Therefore, the conclusions may be a result of culture, not biology. In fact, a couple of body-shape researchers have had the good sense to show the typical "hourglass" drawings of human females to some indigenous Peruvians yet uninfected by the supermodel culture. A common response to these wasp-waisted females was that their bodies looked malnourished, perhaps because they "had diarrhea a few days ago." And other cultures scattered around the world passionately prefer fat females to the hourglass ones. (To date, none prefers underweight females.) So, as a species, we may not carry an innate ideal for the female shape. My prospective mate may have simply compared me to the criteria taught him by his culture.

A more universal, and informative, test was the symmetry analysis we ran on each other, subconsciously. Having lost the details of Homoboy's face in the mists of time, I can at least assess my own and perhaps find the reason we failed to form a lasting pair-bond. Asymmetries are a true signal of troubled time in the womb. They tell on the development of the brain, with a basket of asymmetries corresponding (statistically) with a dip in intelligence. My developmental glitches catch me by surprise when I see a photograph of myself. Accustomed to viewing my face flipped back to me in the mirror, the face I see in photos looks all off-kilter. My asymmetries leap out. To take objective measure of my cockeyedness, I retrieve a carpenter's level from the basement and confront the mirror. With the level across my face I find that the left side of my mandible drops a sixteenth of an inch below the right, giving me a crooked chin. My left eye is a sixteenth lower then my right; my left ear is scary low, off by a strong three-eighths. I'm feeling less intelligent by the second. And it's a head-to-toe problem. My right foot is a good half inch wider than the left, and it's bent.

My esoteric reading on symmetry produces two gems of trivia: I and most humans have a thicker eyebrow on the left than the right, and the grooves on my right fingers are narrower than those on my left. I discover sex-specific imbalances, too. My left mammary gland is a shade larger than my right, as is normal in females; and in males, the right testicle is normally bigger and rides higher.

My sliding features wouldn't be so alarming if they didn't threaten to scuttle my mating prospects. A human confronted with a mating opportunity makes a subconscious survey of the candidate, and even the subtlest asymmetries blunt the attraction. A fun experiment in Jamaica found that females can watch videos of males dancing, and subconsciously pick the most symmetrical ones. Even with the males digitally reduced to faceless animations, the females chose as the best dancers those whom researchers had already determined to be most physically symmetrical. (In the same experiment, the only males who gravitated toward symmetrical females were those who were symmetrically gifted themselves.) This experiment bolsters a theory that music and dance are universal in human culture because they are true signals of mate quality. The world around, dancing is often part of the ritual marking a young human's arrival at breeding age. And judging from the Jamaican study, it seems logical that dance in the human animal could function just like dance in the whooping crane. It draws attention to the performer's exemplary development.

Well, all complicated animals get a bit crooked as they develop in the uterus, with all those cells dividing and redividing. And a few subtle deformities aren't the end of the world. When I check out photos of superbeauties, I find that Brad Pitt's left ear and eye droop like mine, and that Heidi Klum's entire mouth is shifted a half inch left of her nose. And the tension of those glitches probably helps to hold my attention. Every human is lopsided. It's just not great to be the most lopsided. Humans certainly aren't alone in this wobbly world, and we're not the only ones whose most cockeyed players pay a price. A barn swallow with asymmetrical tail feathers is as unlucky in love as is a crooked-faced human. In the final analysis, Homoboy must have passed my symmetry test, and I must have passed his. We moved closer. Sniffing close.

As we saw in chapter 3, humans can glean surprising details from the air flowing off one another. Homoboy and I may have subconsciously scanned each other's immunity profile to see if our portfolios would merge to benefit offspring or would compound our weaknesses. We may have been able to ascertain that we weren't long-lost siblings whose offspring would regret the overlap in our DNA. I guess we passed that test, too. What's next?

To get closer requires a reduction of the human's normal aggressiveness. Humans, although naturally social animals, are nonetheless suspicious of strangers. Unless we're given a reason *not* to, most animals tend to view one another first as competition, and only later as potential friends and mates. Each hummingbird who arrives at my feeder in April regards all others with murderous intent. Males, females, it doesn't matter. Each would happily impale the next with its needle of a beak. It's not until male and female hummingbirds have all chosen individual territories that a female is safe to approach a male and solicit mating. Even so, the male may charge her. Likewise, the solitary female tiger will savage an approaching male even when she's in heat. It may take her days to control her aggression sufficiently so that he can cuddle up without losing an ear. Among mallard ducks, males who fail to curb their aggression are responsible for 10 percent of the female death rate—in subduing their conquests, they drown them. We all have reason to be cautious, it seems. As human pairs dampen their aggression, they signal each other that it's safe to approach. If I know myself, I signaled my friendly intent with copious laughter, arm opening, and head tilting. These are all thought to convey that a human is relaxing her defense of the vulnerable neck and torso. (Researchers have found, conversely, that American females trying to shake off a suitor resort to hand hiding, arm crossing, yawning, close examination of their own hair, and teeth picking.)

And still the tests continued, moving into symbolic gestures. For some species, it would now be time for the male to demonstrate his ability to support the female when she gestates his offspring. These courting gifts aren't common among mammals. Insects and birds make a better showing. A female robber fly must pony up eggs that cost her dearly to create. Shouldn't the male contribute a little nutrition to the cause, in addition to his inexpensive sperm? Many robber fly males approach a female holding a token of insect prey. Male birds ranging from the cardinal to the northern harrier also present prospective mates with seeds or fresh meat, perhaps as evidence of their earning power. The snowy owl has evolved a ritual worthy of prom night, wherein he poses on a mound, wings stretched, and holds in his beak the gift of a dead lemming, dark as a bow tie against his white tuxedo.

In my species, the line between biological and cultural mating behaviors has become cluttered. While gift giving is nearly universal among mating humans, it may nonetheless be cultural, as rituals of pairing often are. But it's possible that the urge to sacrifice is an evolved trait—it could have spread through our species if generosity was a true signal of a male's ability to support a family. Homoboy came up with a photograph of himself, and, if memory isn't mixing up my early prospects, a few bundles of daffodils. In cultures where females are in high demand, gifts can swell to heaps of cattle, cash, or some other "bride price," paid to the female's parents. Where males are at a premium, the economics reverse, and the bride's family makes the gift.

NONSTOP COPPING

Well, there's no avoiding it. It's time for the cop test. While some researchers see copulation as the culmination of the negotiations, others suspect it may be just another way for animals to gauge one another's quality. After all, copulating does not automatically produce offspring. In some species, competing males have ways of killing off one another's sperm in the female reproductive tract; in others, female have ways of rejecting everyone's sperm if they choose. Regardless of whether it's the final exam or the diploma, copping is a big deal.

For most animals, copulation is a serious business. Some insects do it only once in their lives. Most mammals do it only during that slim fraction of time when the female is fertile. Birds may do it many times a day during the week or two that the female is laying eggs, but then abstain for the rest of the year. Tigers, as we've seen, have little desire for one another's company even at the most critical time, let alone for purposes outside of procreation. For most animals, copulation is a chore embraced with the same bosom-heaving passion that I bring to shampooing rugs.

Humans are unusual. For some reason, humans will cop at the drop of a hat. The female's fertility peaks for just a few days each month. But she will cop any day of the week. In fact, the millions of human pairs who practice some form of "rhythm method" of birth control will cop any day *except* those when the female is fertile. This is, biologically speaking, nuts. But for humans, and just a tiny number of other species, copping is distinct from reproduction.

Why? Is it a test of a partner's quality? Some theorists think a roll in the hay might be a good way to gauge another human's health and personality. In my culture it's quite common for a male and female to cop once then never interact again. One or both decides that the other wasn't exactly the right fit. But there is probably more to nonstop copulation than that. My secretive approach to ovulation may also play a part. When a female chimpanzee nears fertility, a "sex skin" on her backside swells like a knobby pink balloon. Everyone in town knows she's preparing to ovulate. This is not my way. When hominids evolved to walk on our hind legs, the female genitalia went into hiding. Even if hominids once advertised their fertility status, the new posture would have blocked the view. The bottom line is I could be ovulating right now, and none would be the wiser. This "cryptic ovulation" presents a problem for the human male. He wants to father my next offspring, but on which day should he solicit mating for best results? One solution is to copulate all month long. Of course, it would help if the female found this an amusing pastime, as opposed to an opportunity for aggression. Which brings us to perhaps the strongest theory.

And that is: When conducted properly, between compatible humans, copulation strengthens the pair-bond. A strong pair-bond in turn benefits the offspring, who enjoy care and feeding from two parents instead of just one. The evolutionary chain of events might have gone something like this: One genetic mutation caused copulation to feel pleasant, even recreational. Another caused the affection hormone oxytocin to rise during copulation, encouraging the copulators to pair up. Offspring of these affectionate copulators thrived under the attention of two parents, and the ensuing dynasty took over the world. It sounds reasonable to me, but there's really no way to prove that's why we're so quick to cop.

If copping were dual-purpose, for both reproduction and bonding, that would help explain homosexual behavior. Like hundreds of other species, the human animal regularly produces homosexual individuals. Somewhere between 1 and 10 percent of humans (more males than females) prefer to copulate with members of the same sex. In some cultures—ancient Greece and Rome most famously—even those males who were mated to females often copulated with other males. (The history of homosexual and bisexual behavior in females is not as well known.) In other cultures, those humans whose homosexuality is pronounced may mate for life with a member of the same sex.

The genetic basis of homosexuality in the human and other animals is increasingly clear. But why would a genetic trait persist in the human population, when the individuals bearing the trait would seem unlikely to reproduce? Some theorize that homosexual humans could be excellent caregivers to their siblings' offspring, some of whom might share the trait. Others point out that homosexuals could be excellent caregivers to their *own* offspring, noting the negotiable (and forcible) quality of human copulation. I prefer the simple, old-fashioned theory: A developing human brain steeps in hormones. More male hormones produce a more male brain. More female hormones produce a more female brain. When the finished brains emerge from the uterus, they range in orientation from red-blooded-male to hot-pink-female, and every shade in between. Will every individual on the spectrum maximize her or his reproductive potential? No. But those who do produce offspring will stir the colors together for another round of full-spectrum brain tinting. Supporting evidence: New research in Italy finds that families with the hottest-pink females—those who bear the most offspring—also had more homosexual males than average. The researchers proposed that the superbreeder females compensate for nonbreeding males, keeping the pink genes in play.

Mysterious as human sexual behavior is, we do have some company in the cop-crazy department. Chimpanzees will sometimes cop outside of the female's fertile window. Dolphins fool around for the fun of it. And, of course, the all-stars of the copulatory circus are the bonobos. This smaller cousin to the chimp is almost always game for a go. Unlike chimpanzees, whose social life is fractious and violent, bonobos rarely encounter a conflict they can't resolve by copulating. When a bonobo group stumbles across a tree full of fruit, a situation that could inspire competitive aggression in other animals, they call time-out and have an orgy. Males and females, young and old, all participate in a grand switcheroo of genital stimulation. Then they eat. But bonobos don't need an excuse to grab a partner and do a little groin rubbing. They'll mix it up just because it's a nice day.

We're no more certain why these animals indulge in non-reproductive cops than we are about our own motivations. In the case of dolphins and bonobos, it can't be a test of potential mates, because in both species individuals copulate in pairings that are guaranteed not to produce offspring. Male dolphins rub and nuzzle with other males. Bonobo juveniles cop with adults, females cop with females, and so on. For the same reason, it's not likely these animals are copping to catch a secret ovulation. Once again it seems to me the bond-building theory fits best for each of the species that behaves as though copulation is more amusing than washing windows.

So it's not surprising that I, like most humans, have occasionally undertaken the act of copulation while forgetting the goal of reproduction. The behavior is educational, revealing a prospect's stamina, intelligence, and communicating ability. And it tickles the brain, the way rich food and addictive drink do. It's just, um, fun.

THE CURSE OF CHEMISTRY

When I encountered my mate-for-life, we passed some of the standardized tests with flying colors. We both displayed kind faces and healthy body shapes. Other tests we skipped, substituting our own. It's good to remember that much mating research is conducted on college students. These are not necessarily mature humans, and many of them lack the experience that helps an older human relegate biological drives to the backseat when choosing a mate. Since about age twenty, my own selection of long-term mates has depended less and less on aesthetics and increasingly on brain-related factors. Over the years I formed bonds with males whose bone structure ranged from high-testosterone to economy, and with facial asymmetry that made flounders look balanced. My mate-for-life has a high-test jaw and shoulders that could heft half a mastodon. His symmetry is serviceable, and his immune system is obnoxiously robust. But I swear he doesn't need these things. What lured this female near was the gray mass beneath his cranial vault. Brain to brain, we pair-bonded.

As pleasant as that connection was, nature had a backup plan. A solid chemical connection linked us, too. Dopamine set a-twinkling a unique constellation of regions in my brain. Dopamine is sometimes called the brain's "reward chemical," because it will twitter in response to good food and sex. In fact, addictions of all kinds catch hold when humans become dependent on the dopamine flush that heralds intoxication. So, there I was in the grips of an addiction-like brain spasm. My onboard opiate system warmed to the task of hooking me to a mate. I was love crazed. Stark raving.

Simultaneously, my testosterone level rose, increasing both my copulatory drive and my energy level. Although my mate was matching me on the dopamine shift, his testosterone level dipped, knocking the corner off his aggression. Whenever we touched, soft bombs of oxytocin, the affection chemical that binds parents to offspring, exploded in our brains. MRI scans have even shown that my brain's centers for critiquing other humans' behavior and intentions were going offline whenever I laid eyes on my newfound mate. In my eyes, he could do no wrong. Love isn't just cuckoo, it's also blind as a naked mole rat.

In my youth, at least, serotonin, another mind-bending chemical, also steered my mating choices. Serotonin is a workhorse brain chemical. It pares away anxiety, smoothes off anger, and probably also removes the aura from around handsome males. But when a human becomes infatuated, the workhorse goes lame, serotonin dries up, and the animal acts addled. Back in the day, when I fixated on a potential mate, obsession was part of the chemical package. And serotonin, or more accurately a serotonin shortage, facilitated the obsessive thinking. This was especially true in the earliest years of my breeding window. Egad, how clearly I recall the newspaper photo of a certain blond lobsterman—lobsterboy, really, or lobsterhominid to keep things biological. His light curls, his deep dimples, the wooden lobster peg clamped between his molars. . . . That scrap of newspaper traveled many miles with me. When he finally came ashore from his natal island for a fishing-related festival, a girlfriend and I craned our necks for a glimpse of the quarry. He was even cuter than in the newspaper. We followed him all day, and then the next day, too. Based on shreds of data, my fevered brain concocted a detailed delusion of who this male was and how we would run into each other's arms across the Town

Landing parking lot, and build a white castle on a hilltop, and make a living raising unicorns. . . .

Even now that I've moved beyond the mortifying effects of low serotonin, the chemical miasma was sufficiently thick to fuse me and my mate-for-life. We proceeded to bond. This was no piddling miracle. Most mammal males and females avoid one another until the ultimatum of reproduction drives them together. Through a narrow crack in the hostility they make contact, copulate, and split. *And don't come back, if you know what's good for you!* The sheer volume of chemicals saturating my brain as I bonded with my mate reveal the extreme measures nature must take in order to get a male and female human to stay in the same shelter for years on end.

This addling of the brain is a sack of dirty tricks, when deployed in a culture that expects a pair-bond to last a lifetime. Most of the chemical effects—the hormone fluctuation, the dopamine, the serotonin—last only a year or two. Then you're suddenly looking across the breakfast table at a deeply flawed and aggravating . . . well, a human. How could you have not noticed that he asks you a question then leaves the room? How could it escape you that she gnaws her nails? And that's why we have oxytocin, which can be renewed daily. As I sit muttering about unanswered questions and unwashed dishes, my mate lays his warm forepaws on my shoulders and kisses my cheek. The oxytocin, always ready to serve, glows in my brain. And we stay bonded for another day.

THE STRUGGLE TO REMAIN BONDED

Even the best chemistry can't settle a human's conflicting impulses to bond and to drive off the competitor. I'm still a territorial animal, after all, leery of yielding control of my space. And I still want the best resources for myself, regardless of how deeply I bury the urge to let my mate spend his precious energy doing all the breadwinning, lawn mowing, car washing, tub scrubbing, and rug shampooing whilst I consume bonbons. And, of course, even the most happily bonded humans still possess the biological and oh-so-adaptive urge to sneak a cop with a stranger. What ensues is a struggle for dominance that provokes growling in some pairs, and mate killing in others.

In cultures where the battlefield is tilted in favor of males, a female takes a big risk in her struggle for dominance. For instance, under the legal codes of Morocco, Jordan, Syria, and Haiti, a male who discovers his mate copulating with another male is partly or completely justified in killing her; and while not legal, such "honor killings" also occur regularly in parts of Pakistan, India, Bangladesh, and a number of other cultures. Females in these groups confront a greater danger of injury any time they challenge their mate.

By contrast, among the gypsy cultures of Pakistan, a female can easily provide for her offspring without male assistance. As a result, both parties of a pair battle with equal determination, and the pair-bond often wears out before combatants do. The same is true in my own culture. The only resource I needed from my mate was companionship. It was solely my decision how much I would squelch my territoriality and self-interest in order to meet that need. To set that sentence in the past tense suggests the war is over. Never. Just this morning I issued the gentlest of growls over his letting hot water—and our shared resources—run down the drain. Just this morning he held eye contact a second longer than usual when I announced my intention to trade resources for a pedicure. The war's never over.

Underlying our civil little war is probably a more primal disagreement. We have agreed to form a pair, but deep down, in our dark, instinctual guts, we both know we would each improve our legacies if we had offspring with a raft of different mates. It doesn't matter that we're both beyond that stage of our lives. Like a fat cat's urge to catch a bird it can't eat, such instincts aren't the sort of thing you can turn off. More problematic, the male and female instincts result in a grand mismatch of agendas.

The human male, if he applies himself, can contribute his genes to hundreds and hundreds of offspring. Genghis Khan, the notorious

Mongol, is presumed to be the ancestor of one in every two hundred humans alive today. As he was subduing Asia, he was apparently subduing a great number of fertile females, too. (This is based on DNA fingerprints shared by about 8 percent of Asian males today; the fingerprint originated about one thousand years ago, and if it isn't Genghis's, historians would like to know whose it is.) The point is, a male's reproductive glory is limited mainly by the number of uteruses he can access. "Big Men" are often best at this, because they can commandeer and provide for many uterus-bearing females. For instance, the late King Sobhuza of Swaziland had seventy wives, estimated to have borne 210 children.

The female capacity, on the other hand, is much smaller. Traditionally, a single pregnancy, with nursing, will engage a female's reproductive gear for three to five years at a time. A male's copulation costs him just fifty calories' worth of exertion and lowers his sperm count for about three days. But for a female, a cop carries the risk of a long moratorium on her breeding activity. It also presents the risk of suffering injury in childbirth, or even death, which catches one in twenty females in some areas. A female therefore chooses her cops carefully. But, as with males, it behooves a female to have offspring with a variety of males. When your environment is unreliable, as it has been for most of human evolution, the best strategy for sending your DNA through the gauntlet of time is to produce a different offspring for every eventuality. The human female, who would be doing very well to raise a total of eight young, one of whose birth might kill her, is picky.

And therein lies the rub. Biologically speaking, both males and females ought to mate with a different partner for each offspring. Biologically speaking, a male's philandering doesn't cost his mate much unless it interferes with his commitment to feed and protect her offspring. But a female's philandering puts her mate at risk of dedicating years to the care and feeding of another male's offspring. This inspires the state we call "jealousy." And females may express just as much of it, perhaps because the risk of a male splitting for good is real and ruinous to her offspring. Thus each partner's attempts to "mate guard" the other is a common source of conflict for a battling pair.

The human female is exquisitely equipped to maximize her breeding options, whether she is pair-bonded or not. New research is showing that she is most discriminating in the days leading up to ovulation. These days are key because she's most likely to conceive if sperm is already waiting when her egg descends. And because sperm lasts a few days in her reproductive tract, she has some time to arrange this. As we saw in chapter 3, the female's senses grow measurably more acute during those days. She's able to see more clearly, hear fainter sounds, and perhaps smell subtler signals of a male's immune profile. Some theorize that this could help her to avoid males who might try to force her to copulate. Others suspect that this is the time when her tests of male quality matter most, and she's sharpening her red pencil. Of course, heightened senses could serve both purposes.

A female's preference for male body type changes at this time, too, whether she's aware of it or not. If you ask the average young Western female to pick a prospective mate from a series of young-male photos, you'll get different answers depending on the time of the month. Most days, she'll favor the softer faces, their bones smoothed by moderate testosterone. Such males are statistically stronger on cooperation and child care. But if she's fertile, chuck all that. Now she wants a rascal and a rogue, a high-rise, high-testosterone male with a cleft chin and a wandering eye. And if you specify that she's shopping for a short-term mate only, the high-test rogues rank even higher.

This is more than academic musing. A mated human female, according to surveys in my culture, is more than twice as likely to seek what biologists call an "extra-pair copulation" during the lead-up to ovulation than during the infertile weeks afterward. Other studies have found females more likely to ditch the mate to spend an evening at a singles bar when they're fertile, too. This is leading theorists to propose that the female human (and perhaps the females of other species) employs a dual strategy regarding males: She pair-bonds with a gentle, generous male who will help with the offspring, and she strives to conceive those offspring with an assortment of dominant, aggressive males. Obviously, this would be a grotesque generalization. But I'd also be a liar if I claimed that the image of some uproariously unsuitable male has never popped into my mind during those hormonally restless days before ovulation. They do come, the parade of rogues, and I'm glad to see them go when the hormones settle.

So, my mate probably had a sound biological motivation to engage in some mate guarding in the early days of our bond. When the subject of other males crept around, his hackles rose and his canine teeth gleamed ferociously. Maybe in my body chemistry he caught a whiff of a cheatin' gene. Fair enough. On a couple of occasions in the past—the way, distant past, I assure you, practically in a previous lifetime—I have darted out of my own territory for an extra-pair cop. I wish I could remember where these cops fell in my fertility cycle, but that data is long gone.

Lots of animals besides humans and chimpanzees have evolved mate-guarding techniques, some of them elaborate. The male gray squirrels in my yard have semen that hardens into a "copulatory plug" in the female's vagina, functioning like a chastity belt. The house cricket includes with his sperm an antiaphrodisiac that deadens a female's libido. Human males, too, have an instinctive reaction to their female's fertility, according to a fresh piece of research. Mated males, shown photos of strange male faces, report stronger animosity toward high-testosterone male faces *when their own female mates are reaching peak fertility*. If this experiment is robust and repeatable, it means that somehow my mate can detect my fertility, and he automatically girds his loins to drive off intruding males.

How would he know, given the human female's cryptic ovulation? At least two studies have found that males find female body odor more "sexy," "pleasant," or "attractive" when a female is approaching ovulation. (These experiments use slept-in T-shirts instead of whole human bodies, so the visual element is not a distraction.) This hints at a subconscious ability in males to sniff out a female's reproductive status. Another experiment found that males (and females, too) are more attracted to photos of female faces taken on fertile days than the same face photographed on nonfertile days. It may be that cryptic ovulation isn't that cryptic after all, and that everyone, albeit subconsciously, is keeping tabs on everyone else's hormonal status.

I was wowed to find a study of female chimpanzees that suggests some of them are as motivated as human females when it comes to selecting the father of their offspring—but they can be much, much more accomplished. Chimps live in large social groups, generally dominated by one male who has the clout (almost) to monopolize a female if he chooses. In one Ivory Coast group, however, researchers checked the DNA of baby chimps to ascertain which males were really fathering which offspring. To their surprise, more than half of the crop was fathered by males from entirely different territories. Female chimps may cop hundreds of times, often with many of their troop's males, during each fertile cycle. But these females were bushwhacking alone to the border, sometimes just for a day or two when they were most fertile, to secure their most crucial cop with a foreign male. Whether these were males they previously knew, or the females were resorting to the equivalent of a singles bar, is not known. But the precision of this phenomenon is superb. The females tolerate their male-dominated world except for the single, solitary day when it matters most.

All this business about female cheating leads to the question of male cheating. A few decades ago, researchers reported that pair-bonded males in the United States claim to form extra-pair liaisons twice as often as pair-bonded females. Up to half of males reported extramarital cops; up to one-quarter of females. But researchers suspect males of exaggerating and females of downplaying. A handful of more recent studies bolster that suspicion, with females appearing just as likely to sneak out of their pair-bond as males are. (I always wondered whom those high-output males were putting out with, if females are so reticent. I'm no mathematician, but I do know copping takes two.) Around the world, the cheating rate varies with culture. In the most male-dominated cultures, females have fewer opportunities to go gene shopping. Even in the United States females who work outside the house are twice as likely to cheat, the speculation being that they have more opportunities than females who stay close to their own territory.

And again, why all this cheating, this nest hopping, this border running? Well, the theory that a human female benefits from an aggressive approach to mate shopping is gaining support from animal research. In birds, at least, cheating often seems to produce the healthiest offspring. The extra-pair exploits of birds are legendary, so they make good subjects. For example, that darling snow goose who mates for life? During the nesting season of the lesser snow goose in Canada, while the female is laying her daily egg the male will scoot away to force himself on some other male's mate. Such conduct is not shocking in bird society. In the nest of your average pair of "monogamous" birds, one egg in ten is fertilized by a strange male. The superb fairy wren of Australia, ostensibly monogamous, is such a superb sneak that the female manages to limit her mate's fertilizations to just one-third of the eggs in their nest. While even one in ten seems high, it may be no higher than the human rate. Evidently nature abhors monogamy.

And probably for good reason. Among cheating birds (which is 86 percent of all species), the offspring that spring from extra-pair cops are often stronger and inherit a more sturdy immune system than their "legitimate" half siblings. Consider the blue tit, a common European songbird. A female tends to cheat with two categories of males: distant ones who are unrelated to her, and close ones who are somewhat related but are large, mature, and talented singers. Despite the evident charms of the large local male, more of the benefits seem to accrue to the less-inbred offspring of the distant male. These offspring are more likely to survive long enough to breed. The male offspring also stand to inherit brighter, bluer crowns that will help them win healthy mates, and the females tend to live longer than their more inbred half sisters. (Why breed with a relative in the first place? This bird's social system may be too stable for much mixing to occur—as also happens in isolated human communities.) Rewards for promiscuity are being discovered in creatures besides birds. Among prairie dogs, the female who cops with just one male will produce fewer offspring than the prairie dog who roams at large. In underfed fruit flies, the female who makes a broad sampling of males has a bigger brood, theoretically because a single underfed male can't provide sufficient sperm for all her eggs.

Cheating has other benefits for females, too. In social animals like chimpanzees and baboons, copping around can protect her offspring from infanticide: A male who has copped with her is less likely to kill her infant, presumably because he remembers that it might be his own. A more devious female motivation might be to deplete the energy and sperm of a particularly promising male, thus denying other females the chance to bear his offspring—who would compete with her own offspring, after all. Whether humans evolved our roving ways for the same reasons, we can only speculate.

Now, why some males have evolved to tolerate the benefit that cheating can convey to females is an interesting question. For many animals, like the prairie dog, the answer is that the male doesn't invest in caring for offspring. So whether a female wants to raise some other male's kids in addition to his is her business. For birds who co-parent, the answer is different. Most birds live such short lives that they may see only one or two breeding seasons. So even if a male catches his mate copping with another male, he probably won't abandon the family. If just one egg in the nest is his, it may nonetheless represent his only chance to reproduce, ever. He's forced to make the best of a bad situation, feeding and defending the whole brood in order to promote his humble legacy. But then again, male birds may be too dumb to catch on to the cheating in the first place. A fascinating survey of birds' brain sizes and cheating rates concluded that those species with the cheatin'est females are also the species with the brainiest females, presumably because those females who can outwit males raise the most fit broods. Only in species where male brains are bigger than females' does the cheating rate dwindle. And, of course, it could be that males are too busy pursuing their own conquests to notice what their partners are up to.

Whether the human animal was meant to pair-bond for life, given our proclivity and the biological pressure to mate promiscuously . . . just isn't clear. One piece of human anatomy argues eloquently that males have evolved to compete with one another for the female uterus. Scientists experimenting with model penises and vaginas, and various recipes for mock sperm, have found an explanation for the strange shape of the male intromittent organ: That cone on the end is ideally shaped to collect and remove fluid from the vagina before making a fresh deposit. And that fluid would be the sperm of a female's previous copulatory partner. Needless to say, an animal who evolved in an atmosphere of stainless monogamy would have no use for such a device. (One of the handsomest such devices belongs to the ebony jewelwing dragonfly. His intromittent organ terminates in two nozzles that snake into a female's twin sperm cisterns. With back-pointing hairs, they scour out the previous suitor's hopes and dreams.)

Does all this cheating affect a human family? How often does a human male succeed in sneaking his offspring into the nest of another male? How often does a female succeed in diversifying the genetics of her brood? In other words, assuming that cheating produces superior offspring in humans, are we all maximizing our potential? The data are poor. Because most paternity testing involves pairs where cheating is already suspected, the sample is unnaturally heavy with cheaters. But among families tested for paternity, as many as 30 percent of offspring are unrelated to the male. Estimates for the larger population range between 2 and 20 percent—in the same range as those most romantic of the mate-for-lifers, the swans.

FAILURE OF THE PAIR-BOND

Given the contentious quality of the human pair-bond, combined with what looks suspiciously like a biological drive to copulate outside that bond, the ties that bind are bound to fray. In my culture, the failure rate has been declining after a steep climb in the 1970s. Young pairs are staying bonded longer, especially if the female has a college education. But busted bonds still proliferate. The venerated American anthropologist Helen Fisher has proposed that perhaps humans, like many migratory birds, evolved to mate for one breeding season at a time. In hunter-gatherer humans, the "nest" empties when an offspring is about four and able to walk independently while its mother forages. And when Fisher analyzed the rate of bond failure in fifty-eight human populations, she found a cluster of decouplings as pairs reach their four-year mark. So perhaps we do have a built-in mating period, which blooms and decays according to a biological timetable.

Ah, don't feel ashamed. Divorce happens even in the best species. About half of all bird species (including the storied emperor penguin) pair-bond only for one breeding season at a time, trying their luck with a fresh partner the following year. Even birds that bond "for life" have been known to call it quits, citing any number of reasons besides irreconcilable differences. Many will split up if they're not producing offspring. The great skua, a predatory seabird, will divorce if an outside female attacks a mated female with sufficient gusto to drive her from the territory. The male, who remains, voices no objection to his revised domestic situation. Likewise for the female mute swan, who has been observed to stand by while an intruder vanquishes her mate, then glide downstream with her new-wedded brute. Flamingo couples almost always split up; masked booby marriages last about half of the time; about 10 percent of mute swan unions dissolve. In our primate family, the "mate for life" white-handed gibbons of Asia have proven to be quite casual about their vows. Scientists have witnessed a young male who successfully serenaded a female away from her older companion, and a female who moved out of her mate's place and spent many months shacked up with a male in another territory before moving back home.

Fortunately, those biodegradable brain chemicals that inspire bonding in the first place are also renewable. Humans can pair-bond repeatedly, even after their reproductive window slams shut. I've done it—formed bonds that lasted for a few years—four times. This is nothing compared to a typical Ache female from Paraguay, who bonds ten times in her life. The nearby Cuna indigenes are also fluid, averaging five bonds each.

THE POINT OF IT ALL: OFFSPRING

Every so often, despite the intersex hostilities that continue right into the female reproductive tract, a sperm cell will make contact with an egg cell. Usually, the blending of the two DNA types goes badly. The uterus detects too many abnormalities in the resulting embryo and ejects it before the mother is aware of its presence. But about one time in four, the male and female DNA harmonizes to form a viable embryo. Or two.

The normal size of a human litter is one. Although one out of eight pregnancies begins with twins, one twin usually vanishes and is absorbed into the mother's body. These are known as "ghost twins." Canadian biologist Scott Forbes, in a gory and entrancing book called *A Natural History of Families*, proposes that humans conceive twins for the same reason that black eagles lay two eggs when they can support only one chick: for insurance purposes. The failure rate for embryos, be they human or eagle, is so high that it pays to make a backup embryo that you'll jettison if the first one works out. Doctors who aid parents with in vitro fertilization do the same, injecting far more embryos than the female could ever grow.

When human twins do occur, they generally share a father but arise from two of the mother's eggs. These fraternal twins can look completely different. But not as different as twins from two eggs and two different fathers. As with cats, birds, wolves, deer, and countless other species, the human female can conceive twins with different fathers. To accomplish this, she must collect the sperm of two males when she's fertile (not a difficult task); she must release two eggs from an ovary (quite common, as we've just seen); then one sperm from each male must find an egg (a little tougher); and both the resulting embryos must be healthy (a lot tougher). But it happens. If it happens to you, please use the correct lingo: *heteropaternal superfecundity.* It's a beautiful mouthful, but the event it describes is so rare in humans that the term rarely has a chance to be pronounced.

Identical twins are more common than heteropaternal twins, and less common than fraternal twins. They are an accident, pure and simple, and reliable the world around. Because they result from the splitting of a single embryo, their shared paternity is never in question. They are, nonetheless, far from identical. As we saw with human symmetry, much can go awry as cells divide and redivide. Each time DNA copies itself, tiny typos, or big ones, can occur. Nine months is a long time to be dividing and copying. It's time enough for an embryo that started life identical to its womb-mate to develop the kind of abnormality—a shortened femur, a massive birthmark, hypothyroidism—that can land a human in the medical journals.

Well, at least we finally have offspring of some kind. I promised in the Introduction to address the question of human hybrids: Can the human animal produce young with any other species, the way a horse and donkey can produce a mule, or a lion and tiger can make a liger? That we know of, the chimpanzee is the only cross that has been attempted. Joseph Stalin conceived the military stratagem of creating warriors that were half human, half chimp. Accordingly in the 1920s Russian artificial-insemination expert Ilya Ivanov injected three female chimps with human sperm (not his own, reportedly). Humpanzees (Or would they be "chimpans"?) were not forthcoming. And just as Ivanonov was lining up five human females willing to receive ape sperm, his last mature male ape died. (He was an orangutan, not that anybody seemed to be feeling picky.) But are three attempts sufficient to rule out the possibility? I would say human hybridization remains an open question.

Once the human female has brought a fetus to term, the next obstacle to reproductive success is the parents themselves. It is surprisingly common for humans to intentionally kill their offspring. Among the strangest accounts I've read was that method practiced by the Ayoreo indigenes of Bolivia and Paraguay. These cultures I encountered in an eerie volume titled *Infanticide.* The female in labor hangs from a tree branch, or perches in the limbs, and lets her infant fall to the ground. If the infant is wanted, the mother's female friends inspect it for deformities without touching it. If it passes, they pick it up. If it fails, or if the family has already decided it cannot afford to feed another infant, the females use a stick to push it into a hole. They bury it.

I'm confident that killing excess offspring is as old as humankind. Until quite recently, it has been the world's most reliable form of family planning. As such, it has been a fixture of human reproduction. For most of our history, females simply haven't been able to feed as many offspring as they conceive. In a subsistence culture, a female cannot carry both a toddler and an infant when she's foraging or migrating to a new camp. Nor can she provide milk for both. Her options, then, have been to abandon the youngest (in which she has invested the least), or keep it and risk starvation of all her offspring and herself. Food stamps are a recent invention. Adoption is, too. Humans evolved in a hand-to-mouth world, where adding one mouth could tip the whole family into crisis. Just like other animals.

Which returns us to that black eagle. The eagle female lays a second egg as an insurance policy, but she lays it a few days after she begins incubating her first egg. Thus, the first egg hatches a few days before the insurance egg. When the second cracks open, the bigger baby takes one look at its puny sibling, and attacks. Over the course of days, it pecks the sibling bloody, then broken, then dead. The parents do not intervene.

And the eagle is no anomaly, no freak. The giant panda is no more sentimental. About half the time, a female bears twins. She selects one and turns her back on the other. The house mouse, humankind's most faithful domestic partner, wastes nothing, including the nutrients incorporated in her brood. If they number too many to feed, well, then they feed her. And there's the noble cardinalfish male, one of the rare fathers to tackle single parenting: A mouth brooder who safeguards eggs in his maw, he'll often . . . swallow. For most animals, reproduction appears to be a black-and-white economic issue: If costs are skyrocketing, it's time to scale back the mission or abort completely.

As for subsistence humans, the reasons they cite for killing offspring often seem bulletproof in their practicality. In summary, these guidelines save infants from a brief and tormented life. Specifically, the most common reasons subsistence humans list for killing new offspring include:

- Resource shortage: If the mother is already nursing an infant, she can't begin nursing another without endangering both.
- Deformities: Without sophisticated medical tools, offspring with developmental or genetic injuries may face a bitter and fruitless trudge through a very short life.
- Twins: They're a challenge even for a mother with every modern convenience at her disposal. For a mother who has to carry her offspring all day while making a living, and nurse them for three to five years, they're often considered impossible.
- Deadbeat father: If the pair who conceived the infant has since unbonded, the infant will frequently be killed. Again, single parenting is a challenge even for a mother with every modern convenience at her disposal.
- Dead mother: If an infant's mother dies in childbirth, her offspring may be killed. In hand-to-mouth cultures, adoption is rarely an option.

These aren't the only reasons. Among the Yanomami, and the Tikopia of Oceania, a male who kidnaps or seduces a female from a competing group will demand that her prior offspring be killed. Copper Eskimos used to expose offspring born in a season when the weather would make it difficult to transport them. Most of the other reasons relate to culture—ritual sacrifices, bad omens, or if the child is female or ugly, for instance. An exception might be the high rate of infanticide among young single mothers. This has been documented in the baby-burying Ayoreo, but may apply to other cultures, including my own. When my newspaper reports on an infant dropped down a garbage chute or pushed into a bathroom trash can, the mother is usually very young. This is a cross-species phenomenon. Among Japanese macaque monkeys, four out of ten first-time mothers abandon their offspring.

As you might expect, all manner of animals abandon their young to certain death for biological reasons. A seal and a bear alike will walk away from her offspring if she's too skinny to provide milk. As we've seen, the panda will abandon one of two twins; so will the California sea otter. It makes no sense for a female to sacrifice her own life for young who would promptly starve without her.

It may be tempting to dismiss human infanticide as a relic of prehistory or a brutish act of ignorance in today's subsistence humans. But it was common in the "civilized" world right through the 1800s. It was only then that England began to combat the practice of overlying, in which an overburdened human pair would present their dead infant to authorities, with the explanation that they accidentally crushed it in their sleep. Parents also routinely abandoned live infants in the streets of London.

As Europe's Christian culture expanded around the globe in recent centuries, the colonizers attempted to implement their own religion's rather fresh-faced ban on infanticide. It didn't always go as planned, because the laws of biology are enforced more strictly than the laws of culture. And the laws of biology dictate that one human body can only produce so much energy to nourish other human bodies. Something has to give. So, for instance, when Christian missionaries imposed their mores on the subsistence cultures of Papua New Guinea, they were successful in reducing the traditional rate of infanticide. But a jump in the rate of infant mortality—infants dying from diseases and starvation—kept the biological books in balance.

In many parts of the world, tools now grant females much finer control over their fecundity. And when sperm does meet egg, and a female finds herself unable to provide for an infant, medical tools allow her to end the pregnancy long before giving birth. The access to all these tools is restricted in some cultures, but the practice of family planning is ancient and the motivation strong. Researchers estimate that humans continue to abort about forty-six million offspring each year, and the practice is roughly as common in cultures where it is banned as where it's legal. And when asked their motivation, a tremendous percentage of females continue to cite a shortage of resources.

A second category of infanticide needs mentioning, unsavory though it may be. This version is practiced by male animals, against offspring not their own. Consider a male chimpanzee who overthrows the alpha chimp of his troop. His next move may seem to lack diplomacy: He'll set about massacring the infants fathered under the former regime. He may harass a mother into a bloody battle, or may await an opportunity to snatch the infant and bite it. The same pattern repeats with a variety of primates, plus a selection of animals as diverse as lions, zebras, grizzly bears, and rats.

A female animal who loses offspring this way will soon stop lactating. This brings her back into a fertile phase. More often than you'd want to believe, she'll copulate with the male who killed her last infant. And why not? This male has demonstrated the fitness necessary to depose a king. His offspring should be strong. And now the female can count on his protection, instead of his persecution.

The question on the tip of my tongue, then, is this: Do human males do the same thing? And the answer is . . . yes-ish. A few major studies have shown that human young who share shelter with a stepfather have an elevated risk of injury and death. The studies are controversial, in part because the subject is so emotional and because only a few cultures have been analyzed. But even the most rabid critic of this research concludes there is a sliver of data that cannot be explained away. My conclusion after reading a hundred pages of research and analysis is that yes, North American males are statistically more likely to injure or kill another male's offspring than they are to kill their own. It's not an epidemic, obviously. And many stepfathers have been shown to invest *more* resources in the unrelated offspring of a current mate than in their own flesh-and-blood offspring from a previous female. But a whiff of harsh reproductive economics still hangs in the air. Female killers are much rarer. In only a few animals—a hamster species, a gerbil, and a lemming among them—do females kill the litters of competing females.

I came into the world weighing nearly ten pounds and caterwauling. For humans, birth traumatizes the infant as well as the mother. I've seen a variety of animals give birth, and it often looks as strenuous as spitting a watermelon seed. But when the human infant's oversized brain encounters the adult female's pelvis, narrowed for running, both parties are at risk. As I squeezed through the bottleneck, the bone plates of my head flexed and deformed. My mother was fully mature and healthy, and we both survived. Another round of reproductive roulette came to a successful conclusion.

Thanks to the migratory pattern of my ancestors, I was born into a tool-wealthy culture. In my American culture, only 25 out of every 1,000 infants born the same year as I was would die of disease, accident, or violence before their first birthday. And as I write this, the number has fallen to 7 per 1,000. This is higher than some cultures, but far lower than most. In Angola nearly 200 of every 1,000 infants perish. In Mozambique it's 131 in 1,000. These high rates are comparable to the rate in chimpanzees, who also lose 1 in 5 infants in their first year. Mountain gorillas fare worse, losing 1 in 4. I made it through the birth canal, and I made it through infancy. Out I stretched, first a bud, then a sprout, then a tiny branch at the tip of the family tree. Up I grew, well fed and watered, and sheltered from killing cold.

When my own reproductive window opened, I experimented like a good primate. But my bonds were tentative and failed time after time. For one thing, my own parents' pair-bond had failed, and that seems to influence the pairing behavior of offspring: They—in the United States at least—tend to pair (marry, in this case) too early, or not marry at all. And when they do formally pair, their bonds fail at twice the normal rate.

Second, I was a female in a gypsylike culture, where I had no need of a male in order to control territory and accumulate resources. My reproductive drive was stuck in idle as I charged hither and yon, running down other goals. From time to time, I'd become aware of that closing window, and I'd vow to have a good think on whether I'd like to reproduce. Then I'd get distracted by global poverty or climate change or the island of Madagascar.

The human animal is turning into an animal that chooses whether or not to fulfill its biological mandate. The task ranks first on the to-do list of every other species. But for a modern human it's optional. In my culture, the number of breeding-age humans choosing not to breed is growing. If this cultural fashion were to catch on worldwide, wouldn't that be an innovative way for an animal to go extinct? Not for lack of habitat or food, but for lack of interest.

8
BUSY AS A BEAVER:
BEHAVIOR

Although *Homo sapiens* is a fundamentally diurnal animal, this, like so many other characteristics, has been modified by tool use and culture. Nonetheless, to a large extent, the primate's activities are restricted to daylight hours. As we have seen, its weak night vision leaves it unsuited for a nocturnal lifestyle. During the human's active period, it gathers food, secures shelter, and cares for young—or trades its labor for the money to purchase those necessities. The daily time budget for this animal varies tremendously among its myriad cultures. In the simplest societies a male might work four hours a day and a female five, with the other hours spent resting, socializing, and playing. But in complex cultures both sexes may work late into the night, leaving little time for other activities. The animal's tremendous reliance on tools both saves time and consumes it, as each new tool in the kit brings its own burden of upkeep.

This is a highly social animal, like the crow, the wolf, the elephant, and, of course, the chimpanzee. Accordingly, its interactions are sophisticated and extensive. The animal uses a tremendous amount of language to facilitate these interactions, even teaching simple phrases to the other species with which it chooses to associate. Despite a reputation for aggression—and males will certainly fight to the death when sufficiently stimulated—humans engage in many more cooperative and pro-social interactions than hostile ones. For instance, acts of altruism are common in this peculiar creature. Only in the most social of other primates does such behavior ever appear, and then in such faint glimmers that it is only now being recognized.

All that being said, the animal is, like so many, a natural xenophobe. It is fearful of strange conspecifics, especially those of a different color. Making it even harder to love, it is status driven, which brings out such competitive behaviors as outlandish acts of self-ornamentation, the defense of oversized territories, and the writing of books.

7:30 a.m.
DIURNAL ANIMAL

I wake for good when the sun brightens the walls of my shelter, two hours later than the cardinal whose territory overlaps mine. He starts his day earlier than I do, reinforcing his boundaries with a wall of sound. Because I extended my yesterday with electric light, I went to sleep later than he did. But at heart, we're both diurnal animals: Up with the sun, down with the sun. That's the animal I rediscover when I ramble a part of the world where electricity is scarce. Under those circumstances, when dusk falls my activities either shrink to the circle of light shed by a candle or campfire, or they cease altogether. Then, come first birdsong, I'm up. Like any good cardinal or capuchin monkey, I use the daylight hours to wrest my food, shelter, and social needs from the environment around me. In the dark hours, when fatigue slows my productivity and a photon shortage shutters my eyes, I sleep to restore my systems.

Prior to the domestication of the photon, *Homos*, with our frail night vision, were held hostage to the sun's timetable. And this worked just fine for millions of years. In fact, the solar schedule still suffices for subsistence humans, although they do stretch their days with the domesticated campfire. In the daylight hours, subsistence humans are able to collect their food, feed their offspring, mend their tools, and still have time left to renew their social bonds. In fact, they have a lot more time for socializing than I have.

My work—the economic version through which I secure my food and shelter—consumes well over half of my waking hours, which nudge deep into the night. Tending to my tools (shoring up my shelter, maintaining my cookware, cleaning my clothing, futzing with my lawn mower) takes another bite. Because my economic work exercises the brain more strenuously than the body, I yield up another chunk of the day to imitation work in a gym. By the end of this typical day, I will have spent less than an hour in social discourse via conversation, e-mail, and the telephone. That's a piddling 7 percent of my waking day. The number bounces up when I share a meal with friends and on days when I converse with another human on my dog walk or at the gym. By the numbers, I'm not as sociable as a baboon (9 to 12 percent) or a bald eagle (12 percent), but I'm more gregarious than a solitary animal like a snake or a bear.

I'd have more time for goofing off if I lived a subsistence lifestyle, especially if I lived a male subsistence lifestyle. A fresh analysis of two subsistence groups in Peru, the Piro and the Machiguenga, offers a typical portrait of human life in the wild: The average adult works less than four hours a day. That's a nice ratio of work to resting and socializing. And digging a little deeper into the data, I find a typical portrait of the division of labor: Piro and Machiguenga combined, an adult male toils an average of 2.3 hours a day; a female does more than twice that. And among the Piro, females work three times longer than males. (I've read oodles of these studies, and I've yet to encounter a culture in which males work more hours than females.) But even the *average* female's lot of five hours is less than half of what

I spend in the trenches. At the end of a week, the put-upon Piro female has worked 45.5 hours (she doesn't take weekends off, but as the owner of an old shelter, neither do I), and I've put in 60 or 70. My Piro sister enjoys 6.4 hours of leisure time *every day*. On my economic work days I get 3 hours—if you count running the dog and going to the gym. On a weekend I do better, often dedicating a solid 6 or 8 hours to the pursuit of happiness. That gives me a weekly average of 27 free hours, compared to my Piro sister's 45. Though it's culturally abnormal, I'm even inclined to spend my vacation time hammering away on the old shelter, instead of recreating. This tally sounds grim. I guess the good news is that my economic labor, if not my daily tool tending, is pretty enjoyable.

I don't know if other creatures enjoy their obligatory tasks. I pondered this recently when "looncam" appeared on the Internet. From my desk I could watch Papa Loon, who took the day shift, sitting on two eggs beside a Maine lake. Sitting. And sitting. And turning his head. And sitting. For twelve hours. If that were my job I'd kick the eggs into the water and drown myself within the first half hour. But the loon brain must have evolved to withstand a burden of tedium that would shatter a human skull. For all we can tell, incubating may be a loon's favorite activity, permitting all manner of mental gymnastics.

The human brain, big and busy, leaps to life each morning and churns with activity all day. In fact, often the brain is still too busy to relax when the body toddles off to the sleeping den. Deep into the night, while the cardinal and the capuchin are dreaming, my brain hums, occupied with tomorrows, yesterdays, and assorted abstractions.

7:45 a.m.
SEEKS COMPANY OF OTHER SPECIES

"Where's your collar, Kuchen?" I ask the canid who shares a shelter with me and my mate. Someday he might learn what this means, but verbs and nouns aren't his thing. This animal, who is about half my weight, bends himself into a C, tail wagging, eyes squinting, and backs into me. We communicate like that: I babble, and he wiggles. It works. Most of what he has to say, after all, translates into "I'm a happy, happy, happy dog!" And what's not to be happy about? While his ancestral wolf sleeps on the hard ground, Kuchen enjoys the innerspring comforts of a mattress. The wolf steps out each morning to stalk and kill breakfast, while Kuchen has only to wiggle in the direction of his dish and it will be filled. Straying outside of his territory, the wolf may find his hide torn to shreds by a hostile neighbor, but Kuchen does his roaming in parks free of predators, and all a-bounce with canine camaraderie. The wolf may live six years before injury or disease pulls him down; Kuchen, whose food arrives even when he's injured, will probably live twice as long. True, he has a smaller brain than the wolf and shorter canine teeth, because the long history his species shares with *H. sapiens* has reduced his need to think and fight. But still, his reward for associating across species looks like a good deal to me.

And my reward for befriending a member of another species? The most immediate benefit I can identify is happiness. Watching such a happy animal makes me grin. But wrapping my arms around him also delivers pleasing sensations of warmth and softness. And the emotional attachment between Kuchen and me is powerfully soothing. I can exercise my gooey oxytocin smoochies on him without the hassle of listening to a list of wants and complaints. And finally, I get a rare and powerful pleasure from making another animal happy, happy, happy. It's a lot harder to do that for a fellow human.

Harder to measure is the effect my dog has on my health. Within my culture, experiments have shown pets to work better than a common medication for controlling stress-related blood pressure. Animals also reduce their owners' stress levels in general, quell the outbursts of Alzheimer's patients, and ease the pain and suffering of hospital patients.

But these are modern benefits. The first domestic dogs didn't eat regularly at human expense, and blood pressure and Alzheimer's are recent worries. Scholars generally propose that *Canis lupus familiaris* and *Homo sapiens* sought each other out of utility. *C. l. familiaris* wanted food, and *H. sapiens* wanted protection and help with the hunting. That's the theory. But I question this whenever I see a photo of a rain-forest urchin clutching a parrot or monkey to his chest. I wouldn't be surprised if the first tamed canids were pups plucked from a den by curious young *H. sapiens.* I can see a string of tykes trailing home clutching these warm, wriggling creatures whose fur was so fun to ruffle. I've done the same thing myself as a teen, extracting a baby raccoon from the roadside weeds where his mother lay dead. Francisco was soon following me around the farm, play fighting with the dog, and trilling in baby talk for food and attention. But only a handful of animals are suited to domestication. When Francisco took to staying out late and raiding the henhouse, my mother drove us both to a distant and suitable habitat and left one of us there. Raccoons cannot be taught docility, let alone a respect for hens. But the list of animals who will live healthy and affectionate lives under human patronage is long. My culture endorses pets ranging from ferrets to rabbits and tropical birds, plus rodents from rats to guinea pigs and hamsters. And from other cultures, both subsistence and modern, I've seen reports of chipmunks, otters, foxes, a variety of monkeys, wild pigs, crickets, sea turtles, and spiders all kept for companionship, as opposed to any other useful feature, including edibility. Our ardor for other species seems limited only by our ability to tolerate the biting and hen stealing.

This is a human peculiarity. Maybe. We know so little about what wild animals do in their private moments that it's probably imprudent to declare all of them strict segregationists. But most examples we have of interspecies affection arise from unnatural circumstances. In a human household, dogs will lie down with cats, and cats will consort with pet mice. In a Tokyo zoo, a rat snake allows a hamster to nap among his coils, while in the San Diego Zoo a lion cub and a puppy become wrestling buddies. Orphaned young animals are particularly prone to cross-species attachments: a captive baby hippo bonds with a giant tortoise; a kitten takes up with a suburban crow; in the wild, an antelope calf cuddles with an ostracized lioness.

But besides these tales of malfunctioning instinct, stories also trickle in about healthy, adult animals making cross-species friends in the wild. One famous polar bear used to visit the outskirts of Churchill, Manitoba, to frolic with a particular sled dog who was chained there. And from a Bolivian wildlife refuge I hear of a wild lion monkey who has taken a shine to a tame spider monkey. The lion monkey braves human proximity to snuggle with the object of its affection. Female wild chimps have been reported to carry around a freshly killed mammal as though it were a rag doll. One adolescent female was seen carrying a dead rock rabbit for fifteen hours, grooming it and sleeping with it. More pragmatic motives drive moray eels to hunt with groupers. Their stalking styles are complementary. Likewise, it's probably not sentiment that spurs colobus and Sykes monkeys to keep company. More likely, the colobus monkeys have learned, or evolved, to exploit the wary and loud-shouting Sykes as excellent, uh, watchdogs.

But I can't imagine that any other species approaches the human degree of fascination with other species. Like an overfed animal in a zoo or a house, humans reach a point where we no longer view other critters as competition. We throw open the door of the shelters we once constructed to keep wildlife out and invite in an ark's worth. We coo at chickens and caress geckos, harbor hedgehogs and pander to pigs, even extend a foolhardy hand to old enemies like cobras, lions, and alligators.

I grew up in a household where neighbors would drop off their abandoned and injured whatnots. My playmates included not just horses and goats, but also a flying squirrel, a chipmunk, and a screech owl. The few years of my young adulthood when I was too migratory to keep a dog were the emptiest of my life. Many humans feel quite out of balance unless there's another species snuffling around the shelter.

7:48 a.m.
COOPERATIVE WITH CONSPECIFICS

Kuchen and I get in the car and head toward the park. I brake gradually at the end of my street so my long-legged boy won't tip off the seat. I wait for another driver to pass, then pull onto Pillsbury. This driving business is an outgrowth of my species' cooperative nature. If you put wolverines behind the wheel, there would be no waiting at stop signs. Animals that live solitary lives don't require one another's trust, so they waste no energy on morality. A wolverine would tear through intersections, savaging any fellow wolverine with the temerity to protest.

Humans, though, cooperate among ourselves. We have to. Throughout our history we've relied on teamwork for hunting animals and finding plant patches, for sentry duty and defense, for mating opportunities, and for building alliances in order to push around other groups of humans. We just couldn't make it as wolverines, wandering alone until mating season drove us temporarily into pairs. And so we have a built-in ability to work together. This isn't a big surprise. Our chimpanzee relatives are cooperative, too. So is a select group of other species, ranging from the two-pound meerkat to the half-ton bottlenose dolphins. Certain lifestyles require cooperative behavior. Ours is one.

The behavior of sharing is so fundamental to human interaction that we do it from dawn till dusk without noticing. Every group of humans that forms a culture forges rules of conduct, then conforms to them, more or less. (More when someone's watching, less when unobserved.) The fundamentals are consistent: Generally, humans in the same group don't kill one another. Usually, group members agree not to copulate with one another's mates or appropriate one another's tools. But the rules can also get pretty baroque (don't chew gum in the subway, don't sell whiskey before noon on Sunday). Regardless, all serve to minimize conflict within a cooperating group. Some groups write the rules down, along with penalties for cheating.

Others, mainly subsistence groups, keep the rules in their heads, and may mete out punishment on a case-by-case basis.

In addition to general peacekeeping behaviors, the human often undertakes specific acts that require coordinated group effort. This can be anything from conducting warfare to building a shelter to trading money for food. Humans are particular about their partners in these efforts. I won't trade twice with someone who takes advantage of me, like the contractor who left my bathroom unfinished. And in my campaign to maintain dog-walking hours on the beach, I collude with dog owners, not dog haters.

Recall that a leading theory on the origin of the huge human brain is that it takes a lot of neurons to keep track of all the favors we owe our allies and the favors we are owed. That same theory has been proposed for the chimp brain and the relatively large brains of crows, dolphins, wolves, and other social animals. Like humans, the chimpanzees also must cooperate in order to maintain their territory, locate food, and defend themselves from bullies in the group. They, too, observe rules of the road, such as "Don't mess with the boss man's favorite female," and "Don't take food from the hand of another." They, too, cheat as often as they can get away with it. And they also choose their allies carefully, according to the task before them. A male hoping to overthrow the alpha will recruit another assertive male, as opposed to teaming up with a shy one. In zoo experiments, chimps have shown that they keep track of which comrades are best at solving a certain puzzle, and choose partners accordingly. They also remember which pal has groomed them or shared food with them. An animal with such a complex social life needs a lot of wattage.

If all this cooperativeness sounds a little too cold and calculated, a little too tit for tat, well, it may be. Even altruism, that gold standard of goodliness, is currently under investigation for posing as something more than a far-sighted brand of self-interest. Altruistic behavior is that which costs me effort, risk, or resources, but which doesn't benefit me. The problem is, it's hard to find an altruistic act that doesn't ultimately strengthen my hand. For instance, when I run into the street to stop traffic for a child who has fallen off his bicycle, I'm benefiting an unrelated child at my own expense. This looks foolish, from a biological perspective. I could get myself splashed out of the gene pool by a Chevy Suburban, while saving the life of a youngster who shares none of my DNA. That sort of behavior just doesn't pay.

But perhaps it does. What if altruism isn't selfless at all, but rather a sly, long-term investment strategy? A flurry of experiments is hammering at this subject now, with results that can be discomfiting to a human accustomed to thinking of herself as a saint. Here's a classic experiment in altruism: A researcher gives me and three other humans each ten dollars. We can keep it all or give some of it to a public fund. In that public fund, every dollar is doubled, and the doubled pot is divided equally among all four players. So, an extreme altruist would give all ten dollars every time, even if no one else gave a cent, leaving her with a five-dollar payback. An extreme "self-oriented" human would give nothing, and still take her portion of the payout. The results of this game vary from culture to culture, with most humans making a substantial investment to the common good, even if the entire game is played over a computer, and players never meet. In short, humans will generally bet a little less than half the ranch that strangers will cooperate.

But if this sounds like good news for innate goodness, hold the kisses. For one thing, it appears that we are hardwired to behave benevolently *when we're being watched.* This was proven by a robot, of all things. Experimenters had already shown that a human's culture impacts how much she'll give, and that American subjects, at least, give more and more as their own identity is made more and more public. Humans have also proven more cooperative when playing with another human than when playing with a computer. But when researchers at Harvard mounted a computer screen beside American players and displayed on that screen the image of a robot named Kismet, who has prominent blue eyes, generosity jumped 29 percent. Because the human brain automatically registers human eyes, the researchers concluded that we may automatically and subconsciously improve our behavior when we're under surveillance. And that would suggest, in turn, that while we might risk some money to test how much our neighbors will cooperate to enrich us, our real generosity emerges only if it's going to enhance our reputations.

Reputation is now strongly suspected as the engine that drives altruism: Because I am such a social animal, it's important to me that other humans trust and respect me. My reputation helps me raise an army of allies when I need them, and also makes *others* more inclined to run into the street in my hour of need. What goes around comes around, in human groups. And most of us try to send around enough assistance to establish ourselves as a solid risk.

Now, perhaps you're thinking, "No, this can't be right. I sent a check to help that little girl who fell down the well, and she never knew it was me." First of all, let's dispense with the girl-down-the-well syndrome. This is the phenomenon in which humans will donate one thousand dollars to aid one human infant, but won't donate one thousand dollars to save one hundred infants in Bangladesh. The difference is that the girl down the well has a reputation. Alas for those one hundred Bangladeshis, their faces and reputations are unknown. So the girl-down-the-well syndrome proves that our goal is not only to save lives. We're up to something else. Even if I run into traffic to save a child and never mention it to a soul, I suspect I'm still up to something besides altruism. That glowing feeling inside gives me away.

The fact that extra-nice feelings well up in a human who acts altruistically suggests that we've evolved to reward ourselves for doing good. How could this come about? Perhaps those ancient hominids whose friendly brain chemistry urged them to cooperate produced more healthy offspring than those who bickered and pilfered from one another. Evolution can work on a group level, if the group members share a feature that helps them outcompete and outbreed the neighboring group. Called "group selection," this mechanism could explain how humans with a built-in urge to be helpful came to rule the world. And *that* could explain why my modern brain still pulses with pleasure when I sacrifice my own welfare to improve someone else's.

What exactly happens to my brain when I hand a PowerBar to a homeless human? For one thing, based on MRI experiments, my trusty dopamine receptors rev up, just as they do for great food, sex, and other life-prolonging goodies. Apparently kindness is addicting. A separate brain region simultaneously dampens my urge for instant gratification, so that I can act in favor of the long-term result. This brain action is notably absent in humans who play an altruism game with a computer. Altruism is strictly a human-to-human thing. When we know our selfish behavior will impact only a microchip, we play for keeps.

Or a monkey-to-monkey thing? Biologists sometimes come across wild animals who seem to take care of one another the way humans might. I've seen reports of a macaque with no hands, and another who was mentally retarded. In both cases, the social group accommodated its needy member with what looked like kindness and understanding. The retarded monkey in particular was permitted to break social rules that would have earned normal monkeys a whack on the head. Apes, too, can look as though they've evolved to enjoy lending a hand. Primatologist Frans de Waal relates the case of a feeble bonobo who was transferred from one zoo to another. Unable to find his way around the chimp enclosure, he wandered the halls in distress until members of his new group returned, took his hand, and led him.

Kindness could also be a cetacean-to-cetacean thing—dolphins and whales are both reputed to stick their noses out to help a comrade in distress. Come to think of it, for thousands of years dolphins have had a reputation, however apocryphal, for aiding distressed humans, too. Quite recently an Australian fisherman credited dolphins with driving away sharks who circled him as he clung to remnants of a shipwreck. Whether such stories are based on fact or fancy is debatable. As is the question of whether the dolphins' great brains would, in such a circumstance, be dancing with dopamine.

So, anyway, here I am driving my dog to the park, cooperating like crazy, behaving like the most benevolent bonobo in the forest. And along comes a wolverine.

I arrive at a four-way intersection with four stop signs. To my right is an elderly male in a red truck. To my left is a youngster in a dark hatchback. The youngster takes her turn and proceeds through the intersection. It's the elderly male's turn. But as he starts forward, a gray-furred female in a silver sedan that was behind the youngster slaps the gas. She doesn't stop, she doesn't yield, she doesn't smile apologetically. This female takes what she wants, which is a left turn in the path of the elderly male. She takes his turn, and she takes my turn, and she takes human cooperation for all it's worth.

Because I have been thinking about this subject, I puzzle over the wolverine's behavior as I follow her to the park. I let my dog out, then bend to her open window and ask the question that was on my mind: "Why would you behave so selfishly just to get here a few seconds earlier?" She gapes in mock bafflement at her passenger and does not reply.

I join my dog and find that my pulse is racing. I have taken a huge social risk. I've punished a noncooperator. Theorists have argued since Darwin over why human niceness persists in spite of cheaters. Theoretically, one cheater should be able to topple the whole system: Silver sedan cheats, which instructs pickup truck that he'll never get ahead just waiting his turn, so he starts cheating, too. Well, if you're both going to cheat, I guess I had better, too! Presto! No peace and harmony!

Punishing is crucial to the survival of cooperation, because punishment erodes the cheater's precious social support. However, punishing also looks like a purely altruistic act: I confront the cheater, and all I get out of it is a racing heart and a peeved wolverine. No dopamine rush, even. Why, then, should I make such a sacrifice for the common good? Once again, the behavior looks biologically bankrupt at first glance. And once again on closer inspection, it appears that punishing cheaters is part of a long-term strategy wherein I trade today's stress for tomorrow's social support. When I volunteer to punish a cheater, I advertise my own high standards for trustworthiness and decency. I attract a better class of allies. My stock rises. Thus goes the theory. I detect, however, a prickle of territoriality in the mix, too. When I approach a dog walker who hasn't scooped her poop, part of what propels me through the anxiety and shyness is a determination to protect my neighborhood from the effect of cheaters. *My* neighborhood, mind you. I am not nearly so quick to confront wolverines outside my realm.

One of the intriguing revelations about punishment is that the more testosterone you have circulating in your blood, the more you're inclined to punish. Furthermore, some males who watch a cheater being punished by electric shock do in fact experience that famous dopamine reward in their brains—watching a cheater get his just deserts can make a male high. Females, though, not only miss out on the dopamine reward but also suffer with the cheater. Neurons dedicated to empathy cause their own brains to feel pain.

Anyway, the sad truth about cooperative behavior seems to be that we're all wolverines inside, wolverines in bonobo clothing. If we consider only the short term, it undeniably serves me best to blow through stop signs, lie to the IRS, and ignore the little girl in the well. But in the long term, I rely on my fellow humans in so many ways that such cheating (in front of them, at least) just doesn't pay.

When a human is punished for cheating, her response is individual, a product of her personality and the culture in which she was raised. Some scarcely seem to notice. But for others, the social sanction is agonizing. In early Icelandic society, the harshest punishment was not death but outlawry, which was reckoned to be worse. The offender was driven from society for a number of years and could be killed on sight by any other human. If he survived his term, he was forgiven. When I visited one of Iceland's putative outlaw hideouts, the horror of this penalty came to life. Some murdering cheater allegedly spent one winter in a hole in the ground about the size of my desk. A spring in one corner provided water. A hide over the top held out the snow. If my memory serves, a frozen horse carcass furnished his meals. And miles and miles and miles of hard, open ground lay between the outlaw and any other human. The lonesomeness would have killed me before the cold and the diet did. Fairy tale though it may be, it's the sort of punishment scenario that strikes fear into the heart of a gregarious hominid.

It would work for me. As a strongly social individual, I'm quick to cooperate, slow to cheat even when no one's watching, and mortified on those occasions when I'm punished. I would make a wretched wolverine.

8:30 a.m.
DISLIKES STRANGERS

When I return to the shelter with my dog, I feed myself and open the newspaper. The first item that grabs my attention is a photograph of a large male. His eyes are closer set than mine, and his hair is darker—what I can see of it under his turban. His nose juts from his face at a steeper angle than mine, and his cheekbones are high and pronounced. His eyebrows are thick. The article says the man is actually a terrorist, but that's not what revs my pulse. Humans are hardwired to fear foreigners. We are, biologically, xenophobes. We fear the unfamiliar *Homo*, especially if he's scowling.

I'll never forget the day I caught sight of myself in a rearview mirror in Madagascar. Having traveled for three weeks without the companionship of a mirror, I had developed an image of humans painted from a warm palette: reddish brown skin, black hair, and mahogany eyes. I had also grown accustomed to reddish brown children shrieking and staring when I walked past their territories. It wasn't until I glimpsed myself in that mirror that the problem jelled. For a moment I didn't recognize myself. All I saw was skin a shocking and diseased shade of pink and hair from which all pigment had faded. And those pale, pale eyes. No wonder children were terrified. I almost shrieked, myself.

If humans are social and cooperative within our group, we're quite impudent with strangers. We appear biologically programmed to expect the worst from any "out-group." It is not surprising. Human groups must defend a territory if we are to eat. We must also defend our access to potential mates, if we are to breed. The upshot is that we probably spent our prehistory the way chimpanzees spend their present: fleeing from strangers if they outnumbered us and crushing them if they didn't.

Xenophobia is not uncommon in the animal world. If you drop a wolf into the territory of another pack, its death will probably be imminent and violent. And heaven help the naked mole rat who tunnels one millimeter too far and finds himself in a neighboring labyrinth. Naked mole rats bite first and ask questions later.

Females sometimes get a break from xenophobia. In many species, it's the young females who disperse out of the territory when they reach breeding age, a behavior that evolved because it prevents incest. The males stay home and await the arrival of females from other territories. Thus, in many animal societies, a pretty young face will be allowed to join the group—after a suitable period of harassment and persecution. But as a human I wouldn't want to bank on this tradition. I've read plenty of accounts of subsistence hunters encountering a strange female on the trail and running her through with a spear. Like mole rats, humans often bite first and ask questions later.

In terms of hard data, a recent study showed how the fear of foreign faces is etched deeply into the human brain. Experimenters presented dark-skinned and light-skinned humans with photos of dark and light faces. Some photos were accompanied by an electric shock. Thus, researchers trained the brains to fear certain faces. Next, they tried to unteach the brains, by displaying those same faces repeatedly, with no shocks. And the brains did indeed let go of their fear—but only of same-race faces. Both races had "shocked" them, but the humans were only able to recover from fear of their own group. Dark-skinned humans maintained their fear response to the pale faces, and vice versa. The exceptions were those humans who had pair-bonded with a human of a different group.

Another attempt to gauge our fear of foreignness took the meta-analysis approach, sifting through 515 studies of human reactions to unfamiliar sexuality, disability, color, or culture. Like the dark-and-light study, this, too, found that the fear faded if the two groups had regular contact. And I suppose that is why xenophobia is a useful adaptation: It prevents me from throwing my arms around unfamiliar humans until I've had a chance to study them awhile. Humans can be very dangerous animals (more on that in a moment), and those ancient hominids who threw the door open to every visitor probably didn't live to warn their offspring about strangers with candy.

9:00 a.m.
ATTRACTION TO SHINY OBJECTS

I commute up a flight of stairs to begin my economic work. My office is newly renovated, and it's a joy to open the door. Yellow paint predominates. The window frame is orange and red.

Halfway around the globe (and the clock), the bowerbird male commutes from his night roost to his bower to resume his own labor of love. It's mating season, and his efforts will translate into a nestful of mini-hims. The flowers he arranged in front of his twig bower yesterday are wilted. He'll need more of those today, perhaps orange instead of pink. And a competitor seems to have stolen half of his shiny black beetles. If he can't steal something to replace them, he'll have to rearrange what's left so that they better complement the white pebbles. Or maybe he'll move the beetles altogether and work with those new red berries he saw. . . .

We are nature's two great artists, the bowerbird and I. The foundation of the bowerbird repertoire is created out of instinct. Each species of bowerbird, for instance, builds a distinct shape from twigs: a maypole, or a tunnel, or an African hut. And each species decorates a different part: the dooryard, the roof, under the eaves. But once a male has built his structure, the final decoration will reflect his unique taste. The arrangement of pink flowers and silver shells, green beetles and blue bottle caps is individual. The application of paint to a bower wall is a personal matter as well, with one bird preferring to work in chewed-up green leaves while another gravitates toward blue berries or black charcoal.

Art in the bowerbird probably evolved as a mating signal: The male with the prettiest bower wins the most attention and the most mating opportunities. Is this why art evolved in humans? And how much of our repertoire is generated purely by instinct, versus individual creativity?

These are murky questions. Many theorists now believe that the impulse to create art—as well as music and a sense of humor—evolved in humans because creativity showcases an individual's intelligence to potential mates. Just as an orange beak in a cardinal advertises his healthy immune system, so my ability to sketch birds and beasts advertises my brilliance. But I'm inclined to believe that my visual shenanigans are less important. I'm more of the school that art is an offshoot of my communication repertoire, or a side effect of having a visually oriented brain and a consciousness that craves self-expression.

We'll never know where art "came from," of course. But it's always gratifying to find someone who shares your personal take on things. That's why, after years of reading complicated theories on art and the lofty meaning of cave pictures, I was delighted to stumble across a theory that takes a great deal of prehistoric rock art down from its pedestal. R. Dale Guthrie, an Alaska biologist, canvassed the rock art of Europe and the United States to determine who, exactly, was making all the handprints and the buffalo paintings. Not to mention the finger squiggles in the mud. And the curiously small footprints in the mud, which are too small, really, to be the tracks of shamans. . . .

Teenage boys, Guthrie concludes. Those were the great artists of the Pleistocene. And most of them weren't so great. The images that don't make it into coffee-table books could be mistaken for my childhood portraits of the horses and deer I saw on our farm. Their heads are too big and their knees bend the wrong way. And like my brother's early works, a surprising number of the animals are jabbed full of spears and spurting blood out their noses. Other themes popular with European cave artists throughout the ages are genitalia, plump females, and plump females copulating with males.

To attribute these visceral doodles to great shamans seems a bit insulting to the great shamans—and to Occam's razor, which is a rule of science stating that your theory should be no more complicated than absolutely necessary. If the simplest explanation for cave art is that boys enjoy both caves and visceral, gory images, then that's your best theory. To argue that adult shamans squirmed into these dark and ungainly locations to make pictures of their hands and of blood-belching deer, for a spiritual purpose to be named later, is a more clattering and encumbered theory. Occam says: Cut it out.

Certainly humans may have produced some rock art for spiritual reasons. And for mating reasons. But those needn't be the reasons I inherited a desire to doodle. I wonder if the human love of imagery might be a "spandrel." Technically an architectural by-product of building an archway, these empty corners have become opportunities for decorative embellishment. The biologist Stephen J. Gould proposed that evolution produces spandrels in animals—useless but harmless features that then become useful. (The more usual course would be that an accidental feature is instantly useful, therefore it's perpetuated.) So perhaps the complex brain of the human animal accidentally, and without fanfare, began to express itself in the sand,

and on stones, and on the skin. Then, because this behavior reflected a brain's sophistry, it began to serve as a mating display. And perhaps the same process generated our sense of humor, our musicality, our propensity to dance, and our urge to tell stories. Humans have enormous "selves," and those selves seem bent on self-expression.

9:10 a.m.
LINGUISTICALLY GYMNASTIC

My economic work involves the manipulation of language. I consume huge amounts of it, written down by others. I produce lesser quantities of my own. To organize my thoughts, before committing language to paper, my brain churns with words. Some of them I write out, shifting them around to assess their interrelationships. The goal of all this wordiness is to pass information along to other humans efficiently. Humans aren't the only animals to use language, but we are, by a landslide, the most articulate, babblative, communicatory, discursive, fluent, garrulous, logorrheic, prolix, verbal, verbose, vociferous, and plain old wordy.

Other animals are lately emerging from the shadows of our ignorance to demonstrate that they, too, use words, if not full-blown grammars. Although humans have great difficulty translating the sounds of most other animals, we are making progress. Bottlenose dolphins seem to have specific sounds they use only as names, for instance. Two dolphins in "conversation" will even appear to mention the name of a third party as they gabble. The putty-nosed monkey has caused a stir among language experts by crying *pyow* to warn of a leopard and *hack* when he spies an eagle. The troop scampers higher or lower in the forest canopy according to which word is shouted. While that sort of word use is common among animals, the monkeys can also combine the two words for a "sentence" that means "Move it!" That kind of manipulation is closer to what humans, whales, and captive great apes can do with words. But so far, humans are the only critter who can manipulate verbiage into an infinite variety of combinations, to yammerize on such themes as our own consciousness and the putative origins of art.

Language lights a fire under another uniquely human capacity, our ability to learn. Of course, tons of other animals can learn. But none shares our enormous capacity for operating in the abstract. If I advise you that refrigeration will make your milk last longer, you can use that information without seeing me put the actual milk in a fridge. But no eavesdropping chimpanzee would profit from that advice.

Chimps, like most animals, learn by observing, not cogitating. A mother chimpanzee fishing termites from a mound does not urge her offspring to gather around and take notes. They just do. (At least the females do. Young males learn termite fishing years later than females, because while the females are watching Mom, the males are jumping on logs and throwing leaves and biting one another's legs.) Again, our knowledge of other species is grossly incomplete, so we should be prepared for surprises—especially from animals whose lifestyle requires complicated behavior. Big cats actively teach cubs how to handle prey by handing over small animals or large ones that are near death. Peregrine falcon parents drop prey birds to their flying offspring to give them practice at hunting on the wing. Meerkats are the latest to demonstrate their educational method for scientists. The diet of these lithe desert mammals includes such lively dishes as spiders and scorpions. Lately scientists have observed adult meerkats going so far as to de-sting a scorpion before giving it to a youngster to kill. But still, not even the clever meerkat can instruct its offspring in the abstractions of algebra.

The human facility with words, combined with a limitless capacity to learn, produces the most complex set of animal behaviors on earth. Building on the efforts of those who came before us, we create culture deep and wide. Aided by culture, we build new tools, invent new terms to describe them, and then generate great swaths of new behavior unrelated to any object in the natural world. Here I am, tapping on plastic squares in order to transfer my thoughts to an electrical imitation of paper, so that I can ponder how the words will register with other humans. This behavior wouldn't serve me terribly well in a world without culture. But my cultural group has agreed that this is the most efficient way to transfer information. We have rearranged our behavior to take advantage of an alphabet, a spelling system, a grammatical tradition, and various tools that speed their production. That's the magic (for better and worse) of culture.

That's not to say that culture is impossible without language. Chimps were the first creatures we observed to create and perpetuate cultural behaviors. Proving that their behavior is more than the product of instinct, different groups of chimps display distinct styles when grooming, for instance. In one region, grooming chimps face each other and raise one arm so they're holding hands overhead while grooming with the free hands. A different chimpanzee group agrees that the best way to groom is with a leaf handy. When you find a parasite, you place it on the leaf for close examination. Then you squish it. The clever cetaceans have also revealed some culture. Female dolphins fishing off the coast of Australia have taken to using a sponge to protect the snout as they scour the seafloor for fish. And their daughters have taken up the custom, so that more than two dozen dolphins were "sponging" as of 2005. (I'm afraid the young males seem to be skipping school.) And at a Buddhist shrine in Thailand where humans feed wild monkeys, those monkeys, but no other population, have added flossing with human hair to their grooming repertoire. So, culture isn't ours alone. But as far as we know, we are alone in erecting such enormous cultural structures as religions, marketplaces, and the World Series. Without language to hasten its spread and evolution, culture will grow rather slowly.

9:30 a.m.

ILLOGICAL, IRRATIONAL, AND COMMITTED TO IT

Firing up my computer, I wonder if this will be one of the days when I break down and peek at my horoscope. It's so seldom instructive. But once in a while, it really strikes a note. . . .

Yes, so my intelligence is magnificent and I'm an expert at exploiting the culture of previous generations. But I'm also an intractable fool. The human brain is vulnerable to the invasion of false beliefs, and once it has gripped one, it's loath to let it go.

Other animals behave in ways that look irrational. But that would imply an ability to think analytically in the first place, a feature nearly nonexistent in nature. Animals can learn behaviors that seem silly—my dog learned a permanent dread of the room in which he once had a bath. But usually, an animal's "silly" behavior is a perfectly sensible behavior around which the context has changed. For instance, a friend once complained that deer insist on crossing highways. I replied that deer don't see it like that. They simply have a nonnegotiable need to travel to their breeding site or their watering hole. This is normal deer instinct. However, when deer produce this healthy behavior in the context of the modern human transportation system, their risks skyrocket.

Humans, on the other hand, behave irrationally in spite of our big brains, perhaps *because* of our big brains. We cling to beliefs that should be destroyed by facts. We leap before we look. We think ourselves into bad decisions. Such loopy behavior seems to be a specialty of only the brightest species.

The most chagrining studies I've read on this subject address the indelible nature of political beliefs. As we saw in chapter 7, the brain regions that conduct critical analysis go dark when we hear something unpleasant about our new mate. Well, the brain behaves similarly with our beloved politicians. In one experiment, researchers popped Democrat and Republican subjects into an MRI machine to monitor their brains. Then they presented negative facts about presidential candidates from both parties and watched those brains writhe. While humans were quick to believe the worst about the opposing candidate, they would not accept bad news about their own man. Brain regions involved in error detection did light up, as though the brain knew it should revise its beliefs. But so did areas dedicated to avoiding the emotional agony of admitting we're wrong. And in many cases, the brains patched together alternate interpretations of the bad news, which then caused regions involved in relief and reward to sparkle. Humans do not like to change our beliefs. In a similar experiment, subjects played a money-trust game like the one described earlier, while in an MRI machine. Before each round subjects were given a description of their single opponent. Some opponents were described as selfish, others as generous, and some were described neutrally. Then all the subjects played, unbeknownst to them, against a computer. Regardless of whether the computer acted selfish or social, the players cleaved to the dogma that their opposite number was munificent or mean. They played accordingly, refusing to alter their behavior even when it proved a losing strategy.

Of course, personality plays a role in most human behaviors, and none more than politics. And personality is strongly influenced by our DNA. One recent investigation sought to pin down the genetic component of human beliefs about a host of hot political topics—abortion, capital punishment, modern art, astrology, property taxes—by studying identical twins. Identical twins share 100 percent of their DNA, while fraternal twins share only 50 percent. So if identical twins share beliefs more often than fraternal twins do, then you could deduce that genes play a part. And do they ever play a part. Using two groups of twins, from the United States and Australia, the investigators concluded that about half the difference between any two humans' ideologies results from genes. Family influence and the social environment account for the other half. This is big news to a species that considers itself autonomous and intellectual. And just as gob smacking as the numbers was the authors' analysis: Humans don't divide neatly into liberals and conservatives, they concluded. Rather, some of us are "contextualists," who tend to be empathetic and tolerant of others, who consider the context before punishing those selfish wolverines, who are suspicious of certainty when they encounter it in others, and who question authority and inequality. Others of us are "absolutists," who prefer a strong group unity with clear leaders, appreciate strict and forceful punishment systems, distrust human nature and outsiders, and are not distressed by inequality. Each of us leans one way or the other, regardless of reason, and these leanings color our behavior.

The seemingly irrational behavior of daredevils may also originate in our DNA. In mice, anyway, a gene called *neuroD2* is known to produce fearless animals who are prone to recklessness. While it's not a foregone conclusion that humans share this gene with mice, it's likely we do. In either case, humans definitely come in the daredevil variety. The speculation is that the seemingly unreasonable behavior of these animals pays off in hard times, when normal behavior won't secure enough food, shelter, or territory to support reproduction. In those instances, the timid humans—or mice or tarantulas—will starve, and the go-getters will go and get. All humans have a touch of the risk taker, and by watching a human's brain in an MRI machine, scientists can predict how each of us will respond to a risky situation: embrace the risk or play it safe. Driven presumably by a combination of genes and experience, two brain regions face off. If the gung ho, carpe diem region glows brighter than the anxiety region, the subject is going to throw caution to the wind and behave recklessly.

Spiritual behavior, famous for its immunity to reason, also appears partially genetic. Using twins again to tease out the genetic influence, a young PhD candidate (and an identical twin herself) has found that identical twins grow up to display a similar degree of spiritual commitment. When the numbers are crunched, it looks as though genes account for 44 percent of the variation in spiritual zeal in a mature human. ("Spirituality" can be expressed as Buddhist meditation, animism, attending worship services, or a veneration of pebbles. It's our impulse to attribute magical powers to ordinary objects or imaginary entities.)

Now, why would genes for clinging to irrational beliefs in the face of stony facts persist in the human population? On the surface, this proclivity hardly seems helpful. But it may illustrate a peculiar "bird in the hand" characteristic of the human brain. A recent investigation of genes and "magical ideation" has granted insight to how a brain resolves a conflict between belief and fact. These researchers focused on belief in the Christian creation myth of Adam and Eve, a classic magical idea. What they discovered is that those humans who are most strongly right-handed are most likely to believe in creation, even when facing mountains of evidence supporting evolution. Why should handedness matter? These researchers theorize that the brain's left hemisphere (which controls the right hand) has the job of maintaining beliefs, and that the right hemisphere (which operates the left hand) monitors the left for beliefs that clash with the observed world. In a human whose right hand does nearly everything, the two brain hemispheres seldom need to communicate. As a result, the right rarely has a chance to survey the left for mistakes. So according to the theory, these super-righties should be among the most stolid defenders of any idea, mythical or otherwise.

As scholars probe our love of constancy, they're discovering an animal who appears to adopt a belief, then chew it like a dog on a bone. Some of us find it easier than others to discard old beliefs if something more mentally tasty comes along. The obvious question is, why this variation? One theorist ventures that in a natural environment that's crawling with predators and opportunities to prey, an ancestral hominid could act fastest if she didn't stop to analyze every event. If she could adopt a general belief, say, that snakes are dangerous, then she needn't waste time analyzing the next snake she stumbles on, but could promptly retreat. Those who did loiter to reconsider their belief were removed from the population. And so humans evolved the capacity to form general beliefs about our world. But, theorizes another theoretician, because we're prone to getting the wrong idea (*all snakes are bad*), it also proved advantageous to update the beliefs on occasion (*some snakes kill only rats*). And we can go overboard in this direction, too. A human who reads one horoscope that rings true and thereupon declares her allegiance to the stars may be updating a little too hastily. And so humans tend to cluster in between the two extremes, some of us going where the wind blows, others stiffening at the whiff of any mental challenge.

It's a bit dispiriting to think that so many immovable suppositions lie just below the surface of my consciousness, shaping my behavior as surely as rocks in a riverbed sculpt the flow of the water.

12:30 p.m.
TOOL JUNKIE

In the middle of the day I break from laboring and seek out food. I cannot count the number of tools I use each day. But the number I use to feed myself might be manageable. Here we have the refrigerator, which itself incorporates a condenser, a motor, a box, a zillion plastic pieces, hinges, gaskets, and an umbilical that connects it to a web of electron-stuffed wires stretching across the continent. Within are glass tools, metal tools, and plastic tools, all holding foods whose route to my shelter involved thousands of additional tools, ranging from seed holders to weed killers, tractors, food washers, conveyor belts, cookers, packages and packing machines, planes, trains, and automobiles, cash registers, and a banking system. From a cupboard holding dozens of ceramic tools I choose a disk on which to assemble my sandwich. I also need that ancient and hallowed helpmeet, a knife, to slice my tomato. And a piece of wood to slice upon. (All of this would be harder and more hazardous if it weren't for the disks of plastic mounted on my eyeballs, bringing the knife and tomato into focus.) Now all I require is a chair and table, plus one sort of paper to clean my fingers, and another with words on it to occupy my brain. . . .

Humans are fanatical tool users. The U.S. Patent and Trademark Office alone approves five hundred new tools each day. Our tools now incorporate dozens, hundreds, thousands of sub-tools. Each NASA space shuttle contains about two million individual parts. If you imagine a day without tools you'll quickly envision yourself naked in the wilderness, hunting down sustenance with your bare hands. I'm afraid you can't even cook your food over a lightning-strike fire unless you have flame-retardant fingers, because the minute you break off a stick for a skewer, you're in toolville. And what did you catch? A toad? An elderly mouse? Hunting is hard, even with tools.

Do other animals use tools? Today, the question sounds ignorant. But not long ago we considered our species the only one with the smarts to handle cutlery. Now we know that chimps use leaves as toilet paper. Capuchin monkeys treat their fur with a spa's worth of medicinal leaves and fruits. Various birds throw stones to break eggs, use moss "sponges" to carry water, steal fish off human fishing lines, deploy bark as a lever, exploit a bewilderment of sticks, grass, pine needles, and bark to pry food from small places, and in the famous case of Betty the captive crow, manufacture a metal hook to extract treats.

So a more interesting question today is, why don't *more* animals use tools? And the answer is probably that they don't need to. Orangutans are a good example. They use tools so rarely in the wild that they were considered unable to until 1993 when one was caught in the act. But in zoos, the red apes are perfectly willing to indulge scientists, using tools as adroitly as chimps, when they're given an incentive. They even have the insight to choose the correct tool today for a job they know they'll confront *tomorrow*. I have a hard time with that task myself.

Well, I do know that I'll want to eat again tomorrow, so I reorganize my tools: plate to sink, tomato to fridge, electric bill to mail so that the cold tool will stay cold. The human tool kit is so vast that much human behavior now concerns the care and feeding of those tools.

3:00 p.m.
AGGRESSIVE

My usual mode of aggression is to snarl at my mate for leaving the jam out with the cover off or to grumble to the Good Neighbors about the Bad Neighbors. The human female isn't often eager to get physical in her aggression. The male is another story.

On my way across the library parking lot I hear the screech of brakes and the squawk of a horn, and discover I'm about to be run over. A scruffy male is now leaning out of his blue car, roaring at me. Alarmed, I point to the crosswalk I'm walking across. I continue toward my car. The brakes screech again and this time the blue metal is at my knee: He has looped around to threaten me again, and is roaring still. I'm glad to be tall and in a busy parking lot. And now I'm feeling pretty aggressive myself. Fixing the male with a challenging stare, I call for backup.

We humans like to exercise our aggression in groups. This is a tradition among the warlike chimpanzees, as well. A male chimp spends a great deal of time grooming his allies so that, when the group's alpha male attacks him, he can call his pals to his side. And his pals will serve. A few minutes after I call my allies, a police cruiser swings into the lot. The officer folds his arms, broadening his silhouette, and waits for my enemy to exit the building. I align myself with him. "Look," my posture says, "when you attack me, you attack my whole clan." And it works. My enemy's posture turns submissive, his shoulders fall, his eyes drop. I lavish vocal grooming on my ally and go about my business.

The male appetite for physical aggression is probably related to testosterone. Recall that a long fourth finger, relative to the second finger, indicates a male's exposure to testosterone in the uterus. And those long fourths are more common in elite athletes and males with autism. Well, that long finger (and presumably the testosterone that produced it) has now admitted a connection to violent behavior. It's not a screaming-loud connection, but statistically, males with those long fourths more often confess a tendency to hit. So testosterone may breed physical aggression. This is far from settled, though. While violent criminals, antisocial males, and impulsive males all tend to show high levels of testosterone in their blood, these are correlations, not causes. It could turn out that impulsiveness causes testosterone to rise, not vice versa. Whatever the reason, the human male is more prone to physical aggression than the human female. That's the case in most other animals, too.

Unlike most animals, though, the human male will occasionally fight until he kills his opponent. In many species, fighting is so risky that the combatants cease the moment one admits defeat and turns away. But a few will press their case until the enemy is dead. Chimpanzees are one. When the males in a troop head toward the territorial border to wage war, they don't intend to throw a ceremonial punch. If they outnumber the neighbors, they will hold down their victims and bite off a fatal number of body parts. Jane Goodall once monitored a troop in Tanzania as it systematically hunted and killed every male in the neighboring territory. (The females they welcomed to join their group. Two out of three did so.) And humans are another exception. Certainly modern tools have made it almost as easy for a human to kill as to injure. But even without the modern tools, humans often behave like chimpanzees, continuing an attack even after their target has ceased to fight back.

The tolerance for hands-on aggression differs widely from one human culture to the next, as we can see from killing rates. The United States, Mexico, and Venezuela share murder rates that are eight to twenty times the rates in Spain and Norway. This doesn't reflect any biological difference between the groups—just cultural variation. In an issue of *Primates*, I read that this seems true of chimpanzee cultures, too. Some groups are trigger-happy, while others rarely kill at all. Some commit most of their murder against neighbors, and others slay their own infants. (I don't mean a male murders his own infant; but rather, the dominant males kill the infants of other males.)

The thesis of this *Primates* paper was even more piquing: It's that hunter-gatherer humans and chimps are equal in their murderousness. According to a grand mash of many studies, both chimps and hunter-gatherers exhibit a rate of murder and war killings of about twice that found in a large American city. When the analysts considered farming societies along with the hunter-gatherers, humans looked even more murderous—three to six times more deadly than chimps. Subsistence farmers are those like the Venezuelan Yanomami and the New Guinea Dani, with Stone Age tools and just a few half-domesticated plants and animals. Unlike hunter-gatherers, farmers are tied to one spot on the land. If they run when violence threatens, they lose their livelihoods. Thus, anthropologists propose, farmers fight harder and die more. In some of these groups, violent aggression claims one in four males.

These murder rates mesh with an old argument that life without a penal code can be extremely hazardous to male health. Take those hard-boiled Tierra del Fuegians we met in chapter 4. Typical of a hunter-gatherer culture, they had no courts or prisons. And, according to E. Lucas Bridges, author of *The Uttermost Part of the Earth*, they killed like crazy. Most often males attacked "out-groups," the neighboring clans, which had offended them by stealing a female or killing a male. But like chimpanzees, they also attacked inside their own groups. Bridges reported that there was no taboo over killing in self-defense or killing one's wife. And the number of females slain for various offenses was notable. Life without a penal code isn't terrific for females, either.

For those who worry that the world is becoming ever more violent, take heart. Consider the previous scenarios as the "before" picture: Males kill one another like chimpanzees, and, depending on their culture, they kidnap and kill females as well. Things have gotten better. Humans still do these things, but not as often.

For the most part, as cultures evolve, they establish rules banning violence. In the "after" portrait, most cultures today exhibit a murder rate that is a shadow of that "one in four males" statistic. Even a big city in the United States might see thirty-five murders per one hundred thousand humans. That's only half as murderous as subsistence-farming cultures like the Dani and the Yanomami.

Before we kill off the subject of human aggressive behavior, I should note that killing is just its most extreme form and a convenient one to tally. We have so many other methods of aggressing. I favor the verbal assault. I'll launch my words head-on, if I'm not worried about getting physically clobbered in return. If I'm unsure of your social strength, I'll verbalize *about* you to anyone who will listen. I'd like to undermine your social support and drive you from the field, without you ever knowing who attacked you. Gossip was once considered a female form of aggression, but now researchers aren't so sure. Males can also conduct a respectable smear campaign and may do so as readily as females.

Once an aggressive act has upset the apple cart, males and females tend to respond in distinct ways. Early research focused on males, as was so often the case. And a male facing danger was deemed to exhibit two reactions: fight or flight. Adrenaline coursed through his veins, whipping up his heart rate and shunting blood to his leg muscles. The blood flow to his skin and digestive tract diminished, and his pupils dilated. Cortisol dumped sugar into his blood to fuel action. Endorphins blunted his sensitivity to pain. Energized and alert, he was ready to attack or run far, far away. This, concluded researchers, is the human way.

Females, we now know, have a different way. A fearful female produces adrenaline and cortisol like a male, but she also serves herself a blast of the love chemical, oxytocin. The same hormone that results from childbirth and copulation mellows the female response to terror. So while a female under fire certainly is primed to fight or flee, she is also equipped to "tend and befriend." And since these bonding behaviors produce additional oxytocin, a stressed female can drug herself out of a panic. In an emergency, she grabs an injured infant, and in ministering to it, resets her chemistry to "calm, cool, and collected." It seems to me the male and female responses complement each other. As a pride of lions circles a human family, the human male inflames himself to conduct an assault or escape, while the female gathers the offspring and the allies.

Aggressive acts are both fearsome and compelling to humans. The facial expressions that register quickest in a human brain are anger and fear. Because acts of aggression put our biological future in jeopardy, they are terribly important and terribly exciting. We often circle around to watch a dangerous confrontation, nervous, but keen to keep track of who's winning.

Fascinating though it is, aggression is not the human animal's most remarkable behavior. More interesting is how *little* aggression humans display, given that we like to live in groups but are territorial at the same time. Sure, I sometimes bark at a political leader I dislike. Recently I curled my lip at an insurance-company representative from whom I've been trying to collect money. And on occasion I will snap at a telemarketer regarding his choice of profession. But my cooperative behaviors outnumber aggressive ones a hundred to one. I don't snarl at the neighbor whose car blocks my driveway. I do trundle across the street in the winter to help another neighbor shovel snow.

Recently a few researchers have investigated the ratio of surliness to sweetness in our fellow primates and found a similar story. Although chimpanzees are famously pugnacious, two surveys found that they spend nearly one-quarter of their time socializing—grooming, playing, and sitting together. The males commit an act of aggression only about every twenty hours, and females every few days. (An act of aggression can be anything from hitting and kicking to swaggering into another primate's space, causing the submissive primate to decamp.) And chimps, along with a few New World monkeys, some lemurs, and the yellow baboon, rank as the most pugilistic of primates. All the rest are far more agreeable and less aggressive.

But that's just us primates. Aggression and affection mingle differently in every creature on Earth. Animals with a more solitary lifestyle than ours (which is most species) rarely feel the love. One raccoon rollicking around my neighborhood in the dark will not chirp cheerfully when he encounters another. A cleaner wrasse staking out a chunk of coral will not pass the time of day with a colleague crossing his territory. The Arctic fox does not join her neighbors for an evening of singing. Compared to many, mine is a strangely sweet and solicitous species.

6:00 p.m.
PLAYFUL ACROSS THE LIFE SPAN

When my economic work has sucked my brains dry and my retinas are branded with the image of the computer screen, it's time to play. Play is a universal element of human behavior, which puts us in a very select club of animals. Although the urge is universal, the opportunities for play vary with human cultures. In my society, a young adult might gather with friends after a long period of work. An elderly one may not work at all, and instead dedicate many hours a day to play. And many breeding-age humans busied with offspring discover that their urge to play remains, but the opportunity has vanished.

Play behavior appears very early in the human. As an infant I could laugh before I could talk. And soon enough most of my daily time budget was devoted to play. Young humans develop so slowly that they're basically useless to their parents. This leaves them free to amuse themselves. It's pretty easy. A mud puddle can kill an hour. Bug chasing consumes another. A stick invested with the personality of a human may prove so engrossing that it's adopted for an entire day or a year. Acorns and horse chestnuts are a crop of riches to be hoarded, traded, rattled, and weaponized.

Play is common among most young animals. The fawn frolics and bounces, the fledgling crow hangs upside down from a branch, the lion cub gnaws on her brother and pounces on her mother's tail. What's all the fooling around good for? It burns calories, so presumably it must accomplish something worthwhile.

The strongest theory is that play is practice for the deadly serious business of survival. The frolicking fawn is rehearsing how to arrange its legs as it flies from a coyote. The crow, too, is practicing dexterity—perhaps for the tactic of grasping another crow's feet and tumbling toward the ground. (Some eagles do this in courting; I've never found an explanation for it in crows.) And the lion's play is all about catching things like frolicking fawns, or fighting off hyenas and competing lions.

But play is probably a multipurpose behavior. It probably teaches me, and other animals, how to "read" another, too: What human hasn't begun shooting hoops or playing poker with a partner, only to discover that the other isn't playing by the rules? Through our childhood games with bullies and cheaters, humans learn to spot troublemakers. Animals apparently use play this way, too. Colorado play scholar and biologist Marc Bekoff observes that coyote pups won't tussle with those peers whom they can't trust to bite gently. He goes on to propose that play lays the foundation for morality in social animals like coyotes and humans, because it rewards animals for treating one another fairly.

Play also prepares a young human for adult roles. This explains the great gulf between the two sexes' play styles. Young females the world over gravitate toward quiet games involving just two or three individuals. They rely on these play partners to meet other social needs, particularly friendship and comfort. This is common behavior among mature females, too: A small group shares child-care duties but also trades the proverbial tea and sympathy. Young males, on the other hand, tend toward what biologists call "rough-and-tumble" play. They travel in big groups, and their games are often goal oriented and fraught with rules. Accordingly, the adult males in subsistence cultures often spend much of their time in groups, joining forces for physically demanding tasks like hunting and warfare. Of course, there's no neat line dividing male and female play styles. Neither my sister nor I ever fixated on a baby doll, and we sometimes enjoyed helping our mother with barn chores. But I did like to play dress-up, and with my best friend, Katie, I could spend an entire day drifting from one make-believe scenario to the next. My brother had a more clear-cut male style of play. He baked cookies, but he added houseflies to the batter. And when Mom dug oily gizmos out of her tractor motor, it was Chip who perched on the tire to help.

The same sexual dichotomy of play is evident in chimpanzees, if not other apes. Young chimp females play gently in small groups or tag along with their mothers, learning useful skills. Young males tear around hollering and savaging one another in jest. But play styles are a dime a dozen. Young gorilla females prefer male playmates and a rough-and-tumble play style. In animals where babies are few, such as solitary deer or a small herd of elephants, the young must improvise. The fawn gambols alone or ricochets off her mother. The baby elephant enlists an adolescent who will kneel to accommodate her. And just as in humans, the occasional tomboy female and gentle male will occur among the chimpanzees, and probably crows, fawns, lions, and elephants, as well.

A young brain is always forging new connections within itself, and each new behavior helps to etch new paths for information. So young animals explore pell-mell, and playing helps them do it. Then as most animals mature, their brains settle into the tried-and-true pathways that have proven most useful in childhood. The brains, and the animals attached to them, become predictable, inflexible, boring. It's a rare day when a grown deer frolics. The mature lioness doesn't pounce upon her neighbor's tail. The crow . . . well, the crow is more like the human, and the chimpanzee, the dolphin, and a few other highly social, intelligent animals: We, the few, the goofy, we play for life.

As I grew, duties ate into playtime. But play persisted. Out went mud-puddle games, but in came dancing and organized sports, painting and music, gossip and joke making. Clear to the end of our lives, humans play. In subsistence cultures chores may be done in an atmosphere of tomfoolery. The world around, a practical joke is always in style. In more tool-dependent societies like mine, the variations of play are endless. Among my own friends, some paint pictures, others play instruments, fiddle with plants, paddle kayaks to an island, motor along the shore to catch fish, motor over to the mall to catch clothes and jewelry, read, play basketball, build furniture, sew quilts, eat at restaurants, take their dogs to a beach, go to movies, got to baseball games, walk up mountains, write novels, knit sweaters, play poker, swim at the pool, lounge in a hot tub, and so on ad infinitum. My dad, a biologist, used to say that the dog has a nearly unlimited capacity for sleep. I would venture that the human has a nearly unlimited capacity for play. My grandmother, who died at one hundred, played card games until the end. She dressed up for Halloween, too.

I could take a page from her book. My brain chemistry lends me a more serious personality, and I struggle to make myself play—if that's not a contradiction in terms. I spend a huge amount of time on economic work, plus dog walking and the fake work at the gym. It's a blessing that preparing food is one of my favorite pastimes. Without that burst of creativity at the end of a workday, I'd be a washout indeed.

7:00 p.m.
ALTERS ITS CONSCIOUSNESS

I cannot pour myself a glass of wine in the evening without worrying. All human brains are vulnerable to addiction because our brains produce such terribly pleasing chemicals when we repeat certain behaviors: sex, eating, various drugs. But some brains are more vulnerable than others. My father was addicted to alcohol for many years of his life. He may have come by that genetically or by training his brain with regular doses of alcohol-induced happiness. If it was genetic, my worrying is warranted.

On the other hand, altering my consciousness is so fun! I've always enjoyed it. The stimulant found in the seed of the North African *Coffee arabica* tree was my first introduction. It giddied me right up, with a springiness of step and quickness of wit that I found I wanted every day. My first sample of the American herb *Nicotiana tabacum* was another revelation. As the nicotine clenched my blood vessels and my oxygen-starved brain went limp, I flopped down in the grass and thought, "That's a nice change from diligence!" Shortly thereafter, a friend and I tried smoking tea leaves, just to see what happened. (We laughed a lot, but that may have had nothing to do with the tea.) And then we smoked *Cannabis sativa*, whose introverting effect was a bad match for my shy personality. *Psilocybe cubensis*, a gray mushroom, was my favorite, producing fabulous hallucinations and daisy chains of Earth-Shaking Thoughts that evaporated by the next morning.

Why do humans—and we do, universally—crave a departure from our normal senses? There's an aspect of this behavior that looks like play. It is vastly amusing, in my opinion, to lie under a pine tree and entertain Grand Ideas as the sun and the needles collude to throw geometric patterns down on one's face. Sadly, hallucinogenic drugs demand more hours than my rigid adulthood can spare. But it is still playful for me to inhale the occasional dose of *Nicotiana* and watch from a distance as my consciousness contracts to a dizzy puddle. And more than occasionally I swallow the fermented remains of fruit or grain, to enjoy the expansive and joyful effect of not worrying about everything on Earth for an hour or two.

But that last, perhaps, that "not worrying," is not so playful. That may represent a second reason that humans crave drugs: to deliver us from the tyranny of our frightfully busy brains.

The human brain is a delicate construction of nerve cells, electricity, and chemicals, all firing and inflating and dampening one another's effects. In the industrialized countries, at least, many of the brains break down at some point in the human life cycle. The human becomes chronically unhappy or *too* happy, or angry, or timid, or unable to control his drive toward food, or sex, or aggression, and so on. The brain just isn't very durable. My own is a typical case. Inside my skull, the serotonin that ferries electrical signals between neurons is in short supply, compared to the average brain. As a result, unless I consume a serotonin booster, I'm a little more anxious than the average *Homo sapiens* and a little more snarly. Both my parents sported this model of brain. The big, fragile organ can go wrong in so many ways. And that probably contributes to the fact that nearly every culture on the planet has devised a method for changing the mental channel.

The exit strategies are almost as numerous as are human cultures. Though I've never tried it, I always thought kava, national beverage of the Pacific island of Vanuatu, was a particularly desperate measure. Traditionally, the kava root's mind-altering kavalactones were activated by chewing a chunk, then hocking it into a communal bowl. The saliva and pulp were mixed with water, strained, and imbibed by all. The recipe is repellent. The flavor is vile. But what a small price for that precious sensation of one's worries vaporizing! I have also not grazed on the leaves of the coca plant. That's the plant of choice for indigenous Peruvians, who somehow divined that adding a pinch of burned seashell powder produced a stimulating effect. And neither have I retched on peyote, the indigenous and upchuck-inducing hallucinogen of the Navajo and other Americans. Its effect on brain chemistry is rumored to be as amiable as that of *Psilocybe cubensis*. No, the strangest drug I've ever consumed was goat vodka.

Humans can (and will) make alcohol from nearly anything: potatoes, barley, sugarcane sludge, rotting fruit, rice, corn, honey, horse milk, camel milk, and yes, goat milk. Making booze is as easy as letting food rot. Leave your moist potatoes or horse milk in the open air, and wild yeasts will fall into it. They'll thank you for the natural sugars and repay you with alcoholic excretions until their own alcoholic waste overwhelms them at a concentration of a few percent. So inventing beerish and wine-ish drinks isn't much of a challenge. Distillation takes more initiative. To produce goat vodka, you have to heat goat "beer" gently enough to vaporize the alcohol without vaporizing all the water and the goat-milk solids. Collect that vapor, cool it off, and you're ready to celebrate. Mongolians have mastered the art. A snort of goat vodka, sipped from a battered aluminum bowl passed hand to hand around the *ger*, captures the pure and fiery essence of a Mongolian Billy.

In keeping with my suspicion that male and female humans are very different animals, I'm curious about how the two react to various drugs. Pharmacology has been slow to recognize the differences between males and females. Most research has been conducted on males. Females are then medicated as though they are small males. But times are changing. For example, recent findings have shown that aspirin saves the two sexes' lives in different ways. In most males, aspirin staves off heart attacks. In the majority of females, though, aspirin prevents strokes. Also in males, opioid painkillers like morphine have a hard time binding to receptors in the brain, so males need more of them. And in females, surgical anesthesia seems to have a similar handicap, leaving females more vulnerable to awakening mid-surgery, but also allowing the average female patient to regain consciousness in about half the time of males.

Is the same true of "recreational" drugs? Of these, scientists know most about nicotine and alcohol, because those are legal, popular, and addicting. Nicotine, like most drugs, causes some desirable changes in the human brain. It sharpens vision and hearing in both sexes, and tunes up mental function, too. In the female brain, it slows the metabolism, which is naturally higher than in the male brain. In males, nicotine ramps up brain metabolism. In a male who runs unusually hot on the hostility, nicotine soothes the savage brain cells—but not in females. Unfortunately, as with many amusing chemicals, there are side effects. Humans addicted to breathing in nicotine risk the ruin of their hearts and lungs, and the lungs of their family members, among a host of other perils.

Alcohol is a marvel for reducing the pace of communication inside the brain, which can be a great relief to the human wearied of that organ's domineering attitude. As a female partial to this sensation,

I'm distressed to note that alcohol-related brain damage, heart disease, and cirrhosis of the liver are more common in my sex than in males. Other sex differences are substantial, too. Females are less likely to become addicted than males. And while three or four drinks dampen a male's brain metabolism by 25 percent, rendering his intellect sluggish, the female brain fights on, burning at just 15 percent below normal. Females stay more alert on alcohol than do males, but report feeling more drunken. The differences, as usual, probably relate to male and female hormones.

Caffeine, which humans extract from tea leaves and coffee seeds, also produces sex-specific effects. It appears to fight Parkinson's disease in males but actually encourages it in females who are using hormone-replacement therapy. And in a curious experiment conducted in the United Kingdom, males who drink a single cup of coffee felt more edgy and anxious, and their heart rates increased; while females felt calm and confident.

Other drugs get less attention, at least of the scientific sort. But through chinks in the scientific wall a few shafts of light are falling on our ancient obsession. Thus illuminated, many recreational drugs are starting to look like medical drugs, a revelation that should not surprise me. *Psiloscybe* mushrooms and the synthetic hallucinogen LSD, for instance, may act as excellent headache remedies. Furthermore, when Johns Hopkins University researchers ran a mushroom experiment recently, the majority of their human mushroom eaters reported that the experience was in the top five most meaningful of their lives. Even months later, the majority reported that the drug had permanently improved their behavior and outlook. Cocaine, we know from its long use among Andean farmers, fights the head-splitting effects of altitude sickness. Opiates like heroin and morphine numb pain when all other drugs fail.

Whether our human need to strum the nervous system is common in other animals is a tough question to answer. Stories abound of tipsy pigs and loopy elephants. But most of them are just that—stories. Without proving that an animal purposely chooses an inebriating food over a nourishing one, all you have is a tipsy critter and some tipsy speculation. A recent paper to flutter across my desk estimated the amount of fermented merula fruit an elephant would have to eat to become legally drunk, and concluded that the long-held view of these animals as determined boozers is unlikely. They'd have to eat four times a normal meal to get a buzz, the researchers calculated. But the scientists also admitted that pachyderms do like their liquor and have been known to suck up rice wine when they discover a stash.

Maybe the best evidence for animals and drug use comes from *Felis cattus*, the domestic cat. The cat clearly seeks out catnip and performs special actions to access the drug. And catnip obviously alters the mental state of a cat. Not so clear-cut but more sinister than a dopehead cat is a cow on locoweed. When other plants run short, a hungry cow will try locoweed. Then she will eat more. And observing this, her friends will try some. The whole crew will grow scruffy, scrawny, and depressed—and continue eating locoweed. But these are domestic animals, with the dullard instincts that domestication produces. I'd like to see some wild-animal junkies before I'm convinced. Accidental stupefaction is probably common among fruit-eating animals, especially those living near humans who farm fruit or ferment beverages. The cases of birds getting zonked on fermented holly berries seem too numerous to be nonsense, but who's to say the birds are eating for the thrill, not just the calories? (Alcohol is extremely caloric.) And while science can easily convert rats and primates to junkies in the lab, that's hardly fair and certainly not natural. So despite the swirling stories, I've yet to see convincing evidence of a wild animal forsaking nutrition and reproduction for inebriation. As yet, scientists working in the wild have turned up not one bacchanalian ape, except for us.

That's not to say animals don't use drugs for other means. We'll see in chapter 10 that our primate cousins around the world treat their fur and innards with a host of plants. Nor does it mean more animals wouldn't take to dissipation if they had the tools. After all, the brain wiring necessary to create addiction is a normal feature in animals. But only humans have harnessed the technology to fire up those reward cells just for fun. And we've created of ourselves a monster. New data show that 37 percent of offspring in the United States live with an adult whose behavior is frequently deranged by one chemical or another, whether that's alcohol, heroin, or methamphetamine. These offspring, as a rule, face an uphill battle to compete with their peers in health and education.

8:00 p.m.
STATUS SEEKING

As darkness falls, I buck the human impulse to crawl under a furry hide and go to sleep. With a sheaf of journal articles and a tamed flame, I curl into a corner of the couch and compete like crazy. I'm not just writing a book here. I'm trying to write a book that's better than any other book. It takes a lot of time, this competing.

Like a zebra or a horse, I keep track of my status in the herd. And like a chickadee, I strive to improve it. And *unlike* a chickadee, whose attainment of high office would win her extra opportunities to raise healthy offspring, I've decided to forgo reproduction. And so I strive for status because . . . ?

Because it's fundamentally human to do so, I suppose. Like eating more than I need and mating with no intention of reproducing, amassing power is a natural urge that looks a tad pointless, in the current, tool-crazed context. I strive because it served my millions of ancestors to strive, and I have inherited their crusading DNA. Whether the contest is for a spot on the bestseller list, or the best-cook award, or the prettiest-female medal, stay outta my way: I'm going for it.

It's a rare animal indeed that doesn't jockey for position. Among my local chickadees, the females make no bones about their goal. In breeding season, each couple is ranked according to the male half's dominance. And if the mate of the first male should perish, the second female will turn her back on her mate in a tail flick, and form a bond with the first male.

But for many critters, perhaps including humans, a position at the top of the social heap brings pain. Among baboons, the boss male is prone to ulcers and heart attacks. The top dog among African wild dogs also has the top rating for stress hormones in his or her blood. (The female dogs are more stressed by high status than are males.) And the mightiest male chimp may also shoulder the heaviest burden of parasites and respiratory disease, an indication that stress is corroding his immune system.

In the human, it's actually those with low status who appear most stressed. Those at the bottom of the status ladder show a high rate of disease, not due to poor medical care or smoking, but apparently due to the misery of feeling inconsequential. Humans, whether singly or in groups, fall to pieces if we think we're not keeping up. In a wonderful little science book called *Happiness* I learned that most of us would rather earn $50,000 in a world where others earn $25,000 than earn $100,000 in a world where others make $250,000. We care more, in other words, about our status than our comfort. And a rare piece of research on human stress hormones found that it's not just money that can sink your status. Among Dominican males those with the highest stress levels weren't the poorest. They were those poorest in social support: They were the social cheaters, the noncooperators, and males who grew up without a father's care. Low status comes in many packages, and each package seems to hold the same unpleasant gift of physical decay. I hope to avoid it. On my couch, I bend again to my pile of research.

11:00 p.m.
SLEEP PERIOD

A time-measuring tool beside my bed counts off the minutes. Dutiful to this device, I roll over and shut my eyes when it says 11:00. Each time I wake in the night, I'll check it. If I haven't slept enough hours away, I'll close my eyes again. If I have, I'll greet the day.

Today the majority of humans sleep thus, wedging a single sleep period between all their other activities. Fewer than half the humans in my culture leave enough time for sleep and are often tired and dopey as a result.

I find the studies of subsistence cultures riveting in this regard. Those humans, from every study I've read, do not lay down their heads and hope for eight hours of oblivion. And they would find my alarm clock an absurdity. Rather, when sleepiness catches up with a human in the middle of the day, she lies down and sleeps. In the evening when stories around the fire lull her, she lies down and sleeps. Awakened in the deep of night by the snap of a twig, she puts wood on the fire and sits for an hour, gazing at the flames. Her brother might wake, too, and together they'll talk about the coming weather or the illness of their mother. As she settles back toward sleep, their cousin may rouse, step away to empty his bladder, then join the brother at the fire. And so the night passes. Humans weave in and out of sleep, punctuating the night with periods of conversation, soft song, or silent watchfulness.

This pattern, combined with the observation that the human's circadian rhythm does not neatly match the twenty-four-hour day of planet Earth, makes me wonder if the human animal didn't evolve to spend some night hours awake, keeping an eye on the world. When I sift through the numbers in a National Sleep Foundation survey, I find another intriguing clue: Three out of ten of my countrymen often find themselves wide awake in the middle of the night. And in a handful of languages, more clues: "Dead sleep" was the Middle English term for the first bout of the night; the second installment was called "morning sleep." The two terms occur in many European languages and at least one Nigerian language, according to American historian A. Roger Ekirch. And like subsistence humans, medieval European farmers also used the wakeful time in the small hours for socializing or solitary reflection. Should I wake tonight before the minute-counter approves it, I'm going to give this further thought.

■ ■ ■

Humans do an awful lot of behaving. As subsistence cultures demonstrate, we can accomplish the essentials in just a few hours a day. But humans are not loons who can sit still for twelve hours of solitary contemplation. That scenario, come to think of it, sounds like the lack of behavior we inflict on criminals. Given our druthers, humans are usually busy—busy building, fixing, playing, or socializing.

Our behavior is in line with that of the other social species, the chimps, the crows, the wolves, and the ants. We cooperate from dawn to dusk. Just today I've disentangled my dog from the neighbor's cat, calmly traded for food, followed the rules of the road, tolerated another neighbor's musical extravagance, greeted strangers on the street, and groomed some friends via e-mail. My aggression I've kept to a minimum, as demanded both by my social biology and my mannerly culture. When I signaled the need for protection from another animal's aggression, help was quick to arrive.

And my daily behavior responds to the extraordinary horsepower contained in my three-pound brain. My reading and writing, the bulk of my communicating, the dance steps I execute in the service of culture, all are behaviors open only to my species. When the loons are lounging by the lake and the lions are lying in the sun, my brain is driving me on, to do, do, do, behave, behave, behave. I am an active ape.

9

CHATTY AS A MAGPIE:

COMMUNICATION

The human animal communicates in typical mammalian fashion, using vocalizations, tonal inflections, physical posturing, and some rhythmic sound production. And as you might expect from an animal with such a powerful social drive and labyrinthine brain, humans warble from dawn till dusk. They attain silence only during the hours of the sleep period, and even then some of them will continue to vocalize.

This animal's repertoire of sounds is rivaled only by that of parrots and a few other birds, and incorporates a huge range of consonant and vowel sounds as well as clicks, clucks, buzzes, and other percussive elements. The human also employs tone shifts—whines, growls, and barks—to convey emotion, in a pattern somewhat interchangeable among many species.

The human's facial lexicon is similar to that of the chimpanzee, incorporating snarls of aggression, fear grimaces, and open-mouthed invitations to socialize. Also like the chimp, the human uses a rhythmic form of communication known as "music" or "drumming." This is often, but not always, conducted in a social group, and presumably serves some important social need, although the precise nature of that benefit is not clear. (In chimpanzees, drumming helps individuals to locate one another, and serves as a territorial announcement.)

Finally, *Homo sapiens* can string together its vocalizations in an innate system known as "language." The power and flexibility of this communication form cannot be overstated. Although language systems are now being recognized in animals as diverse as prairie dogs, humpback whales, and European starlings, human language alone is practically unlimited in its ability to expand to meet the animal's needs. Contrary to popular opinion, the male is equal to the female in noise production. Although the female may not use communication more than the male, some data indicate she uses it with greater facility. Both sexes are enthusiastic liars.

THE HUMAN VOCAL SYSTEM

When I was a blob of cells glued to the wall of my mother's uterus, I undertook a career in communication. I was young and ambitious, eager to get started. Furthermore, I was hungry. The only medium available to me at the time was chemistry. So with an ink of protein, I scrawled missives to my mother: More nourishment, please.

It was a symbolic beginning, in a couple of ways. First, like all animal communication, my signals were bald attempts to bring the world into line with my desires. That theory of communication presents a bleak view of the thousands of words that flow from my face and fingers in a day. But I can't find a flaw in the argument.

And second, my missives kicked off a step-by-step demonstration of how communication may have evolved in the animal kingdom, a reenactment that would unfold as I grew. As mine were, the first acts of communication on Earth were probably chemical telegrams, sent by single-celled organisms. Also early on, animals evolved an ability to sense vibrations: They could hear. Somewhat later, as with my own growth, vocal signals evolved, along with body language. And finally, at least in humans and perhaps in a few other animals, vocalization progressed to verbalizations—sentences, prognostications, riddles. For me, too, this would be the peak on which my communicating abilities would unfurl their flag.

Ah, but for the time being, I was stuck in the womb, equipped only with my chemistry set. I telegraphed a message to my parent: "sFlt1." This particular protein left my body and entered hers. It raised her blood pressure, forcing more nutrition into the placenta for me. *Success!* But her own body, not wanting to have a heart attack for the cause of one uppity offspring, replied with a different protein: "VEGF!" And her blood pressure relaxed. *Rats!* We continued to haggle over what she should sacrifice for me. Although I didn't get everything I wanted, I did talk her out of nine pounds and fourteen ounces, perhaps presaging a career in communication—and a stubborn streak. When these arguments get heated, the mother's blood pressure pounds out of control, and complications like brain swelling, a ruptured liver, and bleeding organs can kill both the mother and her bossy fetus.

This kind of "chem-munication" was pioneered by single-celled organisms (bacteria and such). It allowed one organism to repel or attract another with the release of a few molecules. It remains a popular channel today. The eyeless, earless *Amoeba proteus*, for example, emits a peptide that prevents fellow *proteuses* from cannibalizing it. (And, like eavesdroppers the world over, a delicious ciliated creature called *Euplotes* sniffs this peptide to further its own agenda: It sprouts armor to prevent *proteus* from eating it, too.) With time, evolution built multicellular chemical receptors like noses and tongues. These we humans retain, but they aren't our clearest channel. Our chemical sensors are dull, relative to many other creatures'. They're a ragged connection at best. But when I was a shapeless blob of cells, chemicals were the best I had.

By my sixth month in utero, my vibration detectors were able to sift sound from my watery world. Before this, I could register only loud noises, which made me twitch. But now I began to discriminate one voice from the next. Had someone read a book repeatedly to me at the time, research suggests I would even have learned the rhythm of those words. Then, had someone read me a different book, my heart rate would have hopped, indicating that I was stimulated by the novelty of a new rhythm. My own vocal chords, with no air molecules to knock together, were silent. But I was practicing to be the receiver of vocal signals.

And then, on a dark December morn, my turn to make noise arrived. My lungs, accustomed to circulating amniotic fluid, and then compressed by a trip down the birth canal, took in fresh air. Then expelled it. I went vocal.

It wasn't an eloquent debut. At this stage of my communication career, I operated on a level with the sparrow fledgling or the wolf pup. My vocabulary was not symbolic, but representational. When I was hungry, I didn't say, "I feel peckish." Rather, I inflated the lungs, stretched tight the vocal cords, and squawked. And when I was weary, I didn't announce, "I'm going to hit the hay." I inhaled, tensed, and squawked. And the same for when I experienced gas or a soggy diaper. No conjugation of verbs or checking of tenses for me. Some squawks were distinguishable as cries of pain (they tended to begin abruptly, and last a long time), but for the most part, my complaints sounded alike. And they meant, "I need."

The distress call is in wide circulation among mammals and birds. It's the first line of defense for a young critter in a crisis. The noise is a signal sent to a parent, who receives the message as, "Attend to me!" Once a parent is attracted by this noise, she can research the problem—chill, hunger, fatigue, impending doom—and solve it. When I was in need, I would issue my initial summons at a modest volume. But if no parent materialized in my field of view, I would raise the volume. This is risky, because a distress call can attract predators as well as parents. But because most animal infants are helpless, they're forced to choose between loudly announcing their location and quietly perishing. In my reading on this subject, I'm surprised to learn that some animals can issue the distress call even before they're born. I discover that the embryo of the eared grebe will cheep audibly if she grows too chilly inside her egg. From within the shell, the baby will pipe until a parent applies heat. She'll also speak up to request egg turnings as hatching day approaches, possibly to position her head up and her feet down. In fact, many eggs cheep, I learn, including those of domestic chickens.

How vocal communication first mutated into existence is anybody's guess. The strongest theory is that it all began with some poor infant's attempt to warm its own chilling body in the absence of a parent. Baby rats, voles, and many other small rodents squeak ultrasonically when they're too far from a parent. This tiny noise, some researchers suspect, started as a side effect of something called laryngeal braking. Picture a chilly infant from the dawn of mammal time, who has strayed from his nest and grown cold. By sealing his larynx and pressurizing his lungs, this shivering critter can theoretically force more oxygen into his blood to help burn fat deposits. The burning fat warms him. That's laryngeal braking. Now, if the larynx doesn't seal perfectly, some air is going to squeak out. And if that squeak attracts a parent who then carries the infant back to the nest, well, that leaky, squeaky larynx may just prove to be a blessed mutation. And this, perhaps—the leaking larynx of an ancestral, prodigal shrewlet—is how vocal communication began for all the rest of us mammals.

We have developed bells and whistles aplenty as we've radiated into the rambling class of Mammalia. But all the sound effects derive from one original method, of forcing air through a valve: the lion's roar, the elephant's rumble, the mouse's squeak, the chimpanzee's scream, and the cat's meow. And my own fussing.

For eight or ten months I directed the simplest vocalizations to my parents. Whenever my world fell short of perfection, I squeaked and squawked, and they hastened (or not) to resolve my distress. As receivers of the signal, they could sometimes divine which category of comfort I lacked. Other times, when I was clearly safe and sound, they might try to resist my demands. But the human infant's distress call has evolved to be, as one researcher put it, "noxious." A baby's cry is one of the most grievous sounds to assault a human ear, and for good, evolutionary reason. The hideousness of the noise encourages a quick response. However, just as the wayward mouse pup risks being eaten if she speaks up, so does the human infant who summons the parents too often. Rarely, the signal can be so noxious, and the receiver so overwhelmed, that the receiver responds in a way that harms the infant. My own history was more placid, according to my mother. I was easily appeased. When I wasn't complaining, I was babbling, whining, cooing, and buzzing, practicing a vocal repertoire shared by all human infants under the sun. At ten weeks, I discovered laughter, which Mom says I tossed into the mix at senseless intervals.

Near the end of my first year, I began to assign specific sounds to specific concepts. Instead of "squawk," I managed "Mom." Rather than "squawk," I tried, "Down!" And in lieu of "squawk," I produced, "No." My parents, weary of deciphering my sloppy signals, heaved a sigh of relief. As a practiced receiver, I could already understand a great deal of jabber produced by others. At fourteen months, I could help my sister and mother bake cookies by bringing things they named or opening and shutting cupboards. I was a protolinguist.

Some scholars think my new ability to name things re-created a stage that my primate ancestors passed through on the road to full-fledged language. This stage, known as protolanguage, was a missing link of sorts. It would have been the stepping-stone between the instinctive calls of simpler mammals and the convoluted grammar of modern humans. More sophisticated than an outburst of alarm or anger, this advance in noise making matched one concept with one vocalization. These first "words" would have represented crucial ideas like "come hunting," and "water there." Such vocal symbols would have smoothed the task of coordinating action among hominids. Now they could tell, without the need to show. Although coordinated hunting is feasible without language—African hunting dogs, wolves, and chimpanzees manage it—your clan's kill rate probably improves when you can share a plan in advance.

As language evolved into words for my ancestors, so it was with me. The word "up" now symbolized my desire to be lifted into adult arms. "Go" directed my parents to carry me toward whatever I was looking at. "Me" conveyed my desire for other children's toys. I had no ability to join these words into larger ideas. The words did not relate to one another in my small mind. Each utterance stood alone. But I had tottered into the world of symbols.

All animal brains contain the circuitry to produce communication. What makes a protolinguist like the young me special is an ability to learn. I was born with the ability to distinguish every shade of sound produced in every human language today: some six hundred consonant sounds and two hundred vowel sounds. I was ready to absorb the click language of the !Kung humans, the tonal tongues of east Asia, and the Dutch word for "grow." (It's *groeien.* Try it.) The young beavers and bobcats behind my family's shelter quit their studies after perfecting a handful of instinctive growls and yowls. I was just warming up.

But in the interest of efficiency, my mind's ear soon specialized in the sounds of my own culture. As I committed to hearing all O's as O's, I lost the ability to hear a difference between an Ø and an Å. By just eleven months of age, my linguistic options were narrowing. Still, I lived in a culture that had plenty of terms to teach me. I soaked them up faster than I could pronounce them. Struggling for control of my tongue and the web of muscles around my mouth, I labored to spit out the names of my desires: "bopple" (apple), "mimic" (music), "bite-o" (typewriter).

The mechanics of my mouth are one of the features that make speech so rare in animals. I have very fine control over the muscles, and split-second timing that allows my mouth parts to split the air into syllables, fricatives, and tongue twisters. When anthropologists argue over whether Neanderthals could speak, both sides invoke the necessity of coordinating breath with mouth. A rope of nerve is needed to operate the apparatus, and my own spine's bones have a hole big enough to accommodate such a rope. Is the Neanderthal spine bored wide enough to run a fat nerve to the tongue? Some say aye, others nay.

Neanderthals aside, the animals who can mimic human speech make another interesting comparison. The crows in my yard are able to imitate my voice (though as yet they have demurred), as can European starlings. Parrots, of course, are the masters of mimicry, able to produce a party's worth of chatter. These birds' ability is a case of convergent evolution. The parrots and I don't share an ancient ancestor who could wolf whistle and ask for a cracker. The parrot and I each happened on these abilities independently. And, to the best of human knowledge, which is admittedly poor, only my species evolved to know what it was saying. When Polly squawks that she wants a cracker, she is (probably) issuing a string of sounds that has earned her a reward in the past. But when I, at eight or ten months, learned to envision a cup of white stuff and issue the word "milk," I stepped onto a path trod by a tiny group of animals. I went symbolic.

NONHUMAN VOCAL SYSTEMS

Recently, biologists have made some breakthroughs in translation. Once, we believed each species had just a handful of "words" to work with. Most mammals could say at least "Ow," "Look out!," "Come here," and "Bug off." These noise signals are instinctive in animals, not learned. If you raise a bunch of mouse pups in isolation, shielded from adult instruction, they will still be able to squeak "Ow" and "Bug off." They might even nuance their messages with extra speed or volume, depending on the urgency of the situation. And once upon a time, that was about all biologists expected of animals. Then we began to crack the code of a few species and discovered that some of them are practically poets. Vervet monkeys, it turned out, have three separate warning calls, which translate as "raptor overhead," "leopard underfoot," and "snake." And young vervets don't come into the world preprogrammed with these three symbols. They must learn which noise stands for which predator. Just as I had to learn which noise stands for a glass of milk.

Some biologists complain that each time they discover a new linguistic talent in animals, the linguists invent a new division between animals and humans. At this point, the battery includes five distinctions thought to divide my vocal stylings from those of the animal rabble:

- My words are arbitrary symbols. I can learn to say "nutmeg" when I mean "Let's mate," and as long as my community agrees with the change, I'll be understood. The symbol is arbitrary. Not so with (most) animals.
- I arrange my words in syntactical order. Without the rules of syntax, the meaning falls apart: "I eat the mouse" has meaning. "Mouse I eat the" doesn't. In languages where word order doesn't matter, the form of the individual words changes to clarify how they interrelate.
- I can invent new words, as needed: bungee, snarf, blog.
- My lingo contains small units that combine into bigger groups.
- I can jabber in the abstract, about objects not present and events that have yet to unfold.

It looks like a stiff standard, but various critters are nibbling away at it. Lately I've been reading about prairie dogs, those winsome, squirrel-like burrowers of the North American West. They're making me wonder if humans have a ghost of a clue what we're hearing when the average mammal vocalizes. One researcher has begun to analyze the fine spikes and vibrations that make up a prairie dog's yip or bark. Those little spikes of noise are looking a lot like words. And that makes a yip look like a whole sentence, rather than an instinctual blat of sound.

Like the vervet monkey and a few other animals (some monkeys, some squirrels, and chickens among them), the prairie dog shouts specific words for specific predators—coyotes, dogs, hawks, humans. But these highly social rodents also have calls for animals that aren't their traditional predators: antelope, cows, deer, skunks, badgers, elk, and kitty cats. Prairie dogs are even thought to encode information on the direction an animal approaches from, and at what speed. And when biologist Con Slobodchikoff recently delved deep into the printouts of their chatter, he found the dogs make tiny alterations to their calls in response to such minutia as the color of shirt a nearby researcher is wearing. They even whistle differently in response to differently shaped humans. (To test the latter, Slobodchikoff dressed humans in bulky lab coats to disguise their shapes. The dogs had less to say about them.) This was plenty impressive, since it suggested the rodents were using arbitrary units of sound to symbolize colors and size.

But the prairie dog may be bent on mastering the entire standard for language. To see if the rodents could coin new words, Slobodchikoff presented a few captives with a great horned owl, a predator entirely new to them. Independently, each prairie dog issued the same declaration, a yip Slobodchikoff had never heard. And ditto when they were presented with a foreign ferret. So, can prairie dogs coin new words as I can? Well, the fact that each captive came up with the same word for "owl" suggests that their response may have been more like "Unknown object!" than "Feathers, big eyes, upright." On the other hand, the dogs produced a different term for the ferret than for the owl, so the term was more complex than "Unknown object." And a black plywood oval pulled on a wire through a wild colony solicited yet a third term. So perhaps prairie dogs can coin new words.

Tackling yet another section of the "human language" test, the prairie dogs also appear to use small chunks of sound to build a larger statement. A diagram of the call for "hawk," for instance, contains many of the same peaks and valleys as the call for "coyote," but in a different order and with different repetitions. As for syntax and discussing the future, we just can't tell yet whether prairie dogs are nattering in future perfect. We're too ignorant of their language. And that's what intrigues me about this research: Tons of information sat right under our ears until computer analysis was able to decode it. What further information lurks in those yips and barks is anybody's guess. My own guess is that many other social species are going to prove equally loquacious as researchers learn to analyze the fine print. In fact, just a month after Slobodchikoff published his paper on prairie dogs, another team published work showing that the black-capped chickadee, a bird with a brain the size of a pea, encodes a predator's size in its alarm calls. A few months after that, another team announced that nuthatches eavesdrop on the chickadees, decode their alarm calls, and take shelter according to what they hear. Next it was Siberian jays, whose warnings about predatory hawks specify whether the raptor is perched, circling overhead, or in mid-attack.

Even before Slobodchikoff cracked the prairie-dog code, numerous animals had already taken a bite out of our self-image. Captive chimpanzees have shown that they, like I, can add new words to their lexicon, though they lack the mouthparts to speak them. Even in the wild, chimps apparently augment their natural repertoire with some learned "words." The best example involves the territorial "pant-hoot." The males in each troop of chimpanzees produce a particular pant-hoot pattern that's distinct from the pant-hoot in adjoining territories. This isn't instinctive—the males have to practice and agree on it. That suggests chimps can tinker with an inborn call, adjusting it to sound different from the pant-hoots they hear at their borders. In a similar feat, a handful of birds supplement their instinctive calls with complicated songs they learn from their parents. Most of these birds have to hear their song during a "sensitive window" in the development of their young brains, just as I would have needed to hear Turkish as a child if I were ever to mimic that water ballet of sound. But a few birds, like parrots and the American mockingbird, continue to accumulate words (specifically, bird calls, car alarms, and cell-phone rings) all their lives. And the African vervet monkey has famously combined two of its words ("airborne predator" and "ground predator") to make a new word that apparently means, "Let's all get going!" Whales, too, swap song fragments among themselves, learning and discarding phrases as certain melodies come in and out of style.

And lately the European starling has pecked a mighty hole in one of the strictest distinctions ever to elevate human language over that of the beasts. This is the ability to use "recursive grammar." As animals crept closer to the human standard, linguists added a doozy: Humans can embed one idea inside another. "I talk to prairie dogs" can be set inside "You say I talk to prairie dogs." Or even "Prairie dogs say you say I talk to prairie dogs." Animals can't. *So there!* But now a crew of captive starlings has been taught to use this arcane bit of language. It took up to fifty thousand lessons, and to describe the details of this linguistic accomplishment would take me fifty thousand words, but they can do it. You get the idea. Lots of animals are walking all over the line between human and animal vocal communication. The better we humans become at deciphering animal language, the less unique our own version looks.

In short, humans once considered ourselves utterly unique in our ability to verbalize. Now it looks increasingly as though the linguistic rules I spent years mastering may be learned by a prairie dog pup in a matter of months.

HUMAN POSTURAL COMMUNICATION (BODY TALK)

There I was at a year old, throwing around protolanguage like nobody's business. I was also throwing around my facial features and my hands. My species is equipped with a squishy, maneuverable face. In fact, from day one, I had been studying the flexible faces around me and involuntarily aping them. My mom smiled, and I smiled. She stuck out her tongue, and I did, too. She frowned, and I frowned. These gestures were crucial to master. Seven expressions—happy, sad, surprised, disgusted, afraid, contemptuous, and angry—are shared by every human alive. Along with laughter and tears, they're among the most fundamental tools in the human-communication kit.

At any rate, by ten or twelve months of age, I was not just subconsciously mimicking expressions, I was now decoding them. Suddenly my mother could halt my four-footed explorations with one expression: widened eyes, mouth open, eyebrows raised. At this age, I had learned to check in with a parent when my world presented a puzzle. If I found fear on the adult face, I stopped in my tracks. If I saw happiness, I forged ahead. I was physically literate.

At this stage of development, the difference between my brother and me was probably relevant. When I assessed the emotions displayed on another human face, the visual cortex engaged on both sides of my brain. My brother, viewing the same face, engaged mainly his right visual cortex. The upshot, in the world outside the brain? One of the clearest studies shows that 90 percent of mature females can identify an expression of sadness on an actor's face, while only 40 percent of males can. Scientists also know that females analyze facial expressions faster than males. And some scientists believe I, as a female, alter my breathing rate and posture to match the human I'm communicating with, and that these physical adjustments bring about a matching emotional state in me. This putative phenomenon is called "emotional contagion." It's the emotional version of "yawning contagion." And a flock of studies has indeed found that the female cheek muscles do subconsciously crinkle in response to a happy face and produce a microfrown at an angry face—much more strongly than males'. Another hint about the female receptiveness to physical communication comes from one of those beastly studies in which animals get mild electric shocks. Female humans, it turned out, needed only to watch a comrade get shocked in order for their own brain's pain region to yelp. Females really do feel your pain. And they show it: My female face portrays emotion more clearly than the typical male face, according to tests of both sexes. (Males, on the other hand, can spot an angry face in a crowd quicker than females can. And both sexes identify angry faces quicker than they can identify the less ominous expressions.)

Before we move on to whole-body language, I have to indulge my curiosity in one oddity of face-to-face communication. In my research, I repeatedly stumble on references to one side of my face being more expressive than the other. I dig out photos of myself. My smile is indeed crooked. A lot. And sitting here now, I direct my face to smirk—a lopsided expression by definition—and the left side of my mouth gets the action every time. Same result for a deliberately insincere smile. Weirder still, this subject came to scientists' attention when someone noticed that formal portraits have traditionally been painted to show more of the left side of a face. Experiments soon demonstrated that when you ask humans to display an emotion, they turn the left cheek toward you, while those instructed to show no emotion tend to present the right. What's up here?

No one knows yet, but it is a scientific fact that humans, male and female alike, have an easier time reading emotions that are displayed on the left side of the face. One way to test this is to manufacture two faces from one photo of an expressive face, using two right sides to make one face, and two lefts for the other. And sure enough, humans can read emotions more easily on a left-left face than a right-right face. Peculiar, eh? Naturally there are exceptions, because humans are confoundingly complicated. It looks as though negative emotions (sadness and anger) show up better on the left, whereas happiness spreads more evenly across both sides. And sometimes sadness pops up on the right side, if the emotion is particularly weak or strong. And one study (just one, mind you) found that Japanese humans display their negative emotions on the right side, rattling the whole question of whether this effect is biological or cultural. Argh! Humans! Obviously we have a rich repertoire of physical communication. But our propensity to create complicated cultures makes it tough to write a simple dictionary of which bodily expressions we inherited from biology, and which we learned from our friends.

NONHUMAN BODY TALK

My childhood talent for body language drew on a long history of animals groveling and baring their teeth at one another in a fairly straightforward style. Even the world's lizards and snakes, most of whom are silent save for a clench-throated hiss, are vivid gesticulators. The common garter snakes of my youth lacked vocal cords but signaled clearly with their fangs when I tried their patience. I've met chameleons the size of my thumb who, when my thumb approached them, fixed both eyes upon me and gnashed their microscopic dentition. I've met other chameleons whose very hide flashed chartreuse and violet to convey the rage that my approach provoked. Even a threatened spider hanging in the garden sends a message—a lie, no less—when she bounces in her web, portraying herself as larger than life. And, of course, the more social and big brained the species, the more extensive the animal's bodily vocabulary.

Chimpanzees, for instance, aren't gifted orators. Their calls seem to be few, and some of them, like the pant-hoot, they deploy for a lot of different situations. But chimps are regular blabbermouths with the body language. It's so rich because like me, chimps start with a foundation of instinctive gestures, then add gestures they learn or invent.

Many of the chimps' basic gestures are similar to mine. A chimpanzee's "play face" compares well to my smile: The mouth is open a bit, and the bottom teeth are bared. The chimp's angry face is also like mine: The lips are tightly compressed. And a chimp's disappointed or anxious face corresponds to my "disgusted" face: lips pursed as though to say "Eww." And that's just the face. When a dominant chimp is approached by an underling he doesn't care to see, he may hold up a very human-looking "stop" hand, or even wave off the creature with the back of his hand. If he does want a chimp to approach, he reaches out and flexes his fingers in a "come here" that looks exactly like mine. If he wants a piece of fruit, he'll tender a begging hand. Our groveling maneuvers are similar, too. I'm likely to tilt my head disarmingly and open my arms when I'm appealing to someone in a dominant position (a parking-ticket guy, for instance). A chimp in a similar situation also tries to appear unthreatening, by bowing or leaning away. A female might present her backside for a symbolic mating. (For the record, I have never done this, even for a traffic officer.) Furthermore, chimps use pats, hugs, and kisses to reassure one another. And their grooming might be a variation on the hand holding, back rubbing, and hair braiding I trade within my own group.

Human and chimp gestures don't all translate directly. A chimp's toothy grin, exposing uppers and lowers, is a fear grimace, usually deployed toward a brute who's coming to beat up the grimacer. A different "smile," this one with all teeth covered, is issued by the approaching brute. It doesn't signal happiness. And because both our species are creative, chimp and human gestures diverge wildly once outside the range of basic communications.

One of my favorite chimp signals, a cultural invention particular to one wild troop, is "leaf clipping." When a fertile female is around, a randy male might eye her, display his penis, and then rattle a leaf in his mouth. There's something endearingly goofy about this, something just as ludicrous as the wolf whistle of the human male. In another wild troop, male chimps try to catch a female's attention by rapping their knuckles on a small tree. Again, should the female have any question about the male's intent, he displays his penis at the same time. A male will also shake leafy branches at the object of his affection. Juveniles have a lingo of their own. A youngster who wants to rouse another to play might slap the ground, poke or throw sticks at his target, bite a branch, flop on his back, or even leaf clip. And, of course, the top males in a troop share jargon unique to their position. They drum rhythmically on the flaring buttresses of trees to telegraph their locations. When two rivals meet they may rise on two legs, thrust out their elbows, and swagger like human thugs to intimidate each other. (Their hair stands up, too, to make them artificially huge.) A male in the full grip of a testosterone fit will careen around the forest like a two-year-old in a tantrum, breaking branches to drag around, hurling rocks, yanking up saplings by the roots, and slapping the ground. Females are more restrained, generally initiating communication when their offspring need comforting, or to whack away another female, or to signal their allegiance in a conflict by quietly moving to sit beside an ally. (The bonobos, sibling species to the chimpanzee, use many of the same basic gestures, but also communicate their apologies, reassurance, forgiveness, fondness, and a dictionary's worth of other messages via the language of sexual gymnastics.)

And that's just the chimps. Some spiders send their physical telegrams by scratching the ground or plucking web strings. Various fish can click their teeth, play their swim bladders like bagpipes, or strum body parts, as do crickets and cicadas. Zebras, pumas, rabbits, and rats all lay back their ears when they're fixing to nail someone. Even deep-sea jelly creatures will light up like angry little Ferris wheels if they detect a disturbance in the ocean around them, a spectacle I encountered while diving deep in the *Alvin* submarine. As our little craft sank through herds of these organisms, their responses punctuated the perfect blackness of the water. We all have something to communicate, and if we can't tell it, we'll show it.

EVOLUTION, AS ILLUMINATED BY THEVOCALIZATION OF HUMAN YOUNG

At eighteen months, I was able to string together words to convey rather complicated information. "Sit in lap," I would demand when I wanted cuddling. Or, "Go see horse today jump over stick?" when I hoped for a barn visit. That's not to say I had the system completely under control. I might announce, as it is recorded in family correspondence, "Sometime Daddy be right back soon maybe." For a while I felt strongly that the zippers in Mom's sewing basket were called "rabbits." But I was largely making myself understood. I was sending and receiving signals like a pro.

And yet, ringing in my memory are two sentences I read recently, pronouncements of Peter Gärdenfors, a cognitive scientist in Sweden. He was referring to animals in general, including me:

"All communication is a sign of failure. If everybody is pleased with the situation, then there is no need for communication."

Wow. It unnerves me to think of all my daily words in that context. However, it's true that my earliest communications were all about the shortcomings of my environment. And as I matured, I suppose my clattering has only become more subtle in its pursuit of a planet arranged to suit me best. Case in point: I'd like this brick of communication in your paws to help you understand better your position in the web of life, and perhaps shift your behavior to safeguard the other living things that I personally hold dear.

So as I grew, and my brain soaked in words, I encountered failure upon failure in my world. I had a great deal to say. Often all I wanted was to see eyes turn and rest on me. Other times I needed information to help operate the world around me: "What's that?" Or to test the power of a simple question to extract noise from an adult: Why? "Because dogs sometimes eat their vomit, honey." Why? "Because . . . they think it's food." Why? "Umm . . . because they're dogs." Why?

As my vocabulary expanded, my voice deepened. Had I been growing up much earlier in the evolution of animals, my sinking voice would have been a direct reflection of my expanding size. That's still the case in many animals, due to simple physics. As a rule of thumb, a mouse or an alligator is hard-pressed to generate a sound wave longer than its body. So a small gator has a higher voice than a giant gator. But the depth of my voice isn't so straightforward. Size does play a part, because the human throat tends to be longer in tall humans. The throat acts as an echo chamber, and a longer chamber can produce sound waves that are lower and longer. But the size of the vibrating flaps in my larynx also determine how low or high I can go. It's possible for a human to have a low-slung larynx fitted with high-strung vocal cords. So the human voice can end up anywhere on the spectrum.

As I grew, my throat got longer and my voice got bigger. But my larynx also did a uniquely human thing: It shifted down, down, down, all on its own. In most mammals, the spine joins the skull at the rear. The throat and trachea run parallel to the spine, straight back toward the gut. But as hominids evolved into upright animals, the spine swung down until it was poking into the skull from the bottom. To make room, the throat was forced to take a dive, too. In this remodeled anatomy, the larynx found room for itself halfway down the neck. And in each growing human, it actually migrates there. My brother's larynx, propelled by testosterone in his adolescence, sank even lower than mine, converting him to a rumbly baritone. This issue of bigger animals and deeper noises brings us to another theory on how vocal communication got started in the first place. This one originates way back in the reptilian days.

If you didn't care for the laryngeal breaking theory, try this: Vocalization caught on because yelling proved a more successful way to fight than hitting. This theory proposes that it all began with the reptiles and amphibians, animals that continue to get bigger and bigger all their lives. To appreciate this theory, we have to start with the proposition that these animals were not naturally sociable, like humans. Their dealings with one another were usually competitive and not at all polite. The primary message they desired to send was "Come closer and I'll bite you." Before they evolved a way to send such a signal, they were forced to settle their disputes by actually biting one another. That was the only way to determine who was bigger and thus entitled to respect. But then one ancient animal ancestor was born with a defect that allowed his airway to vibrate like a bad muffler. This innovation allowed animals to fight their battles from a distance. Rather than walking right up to his enemy, a vocal animal could vibrate his throat and telegraph his size through the air. This represented a tremendous savings in blood, infections, and lost limbs.

Those species who inherited an ability to wage remote warfare lived longer, healthier lives, and passed the noisy windpipe on down the line. That's that theory: Vocal communication became a smashing success in the animal kingdom because it cut down on biting.

For doubters, one more theory: Time was, a speechless mother had no way to warn her offspring about the approach of a predator. She could only stand by dumbly as her legacy was mauled. Then a noisy-breathing mutation struck. When this malformed mother spotted a predator, her breath rushed through her misshapen throat with a noise like leaves in the wind. Her progeny took heed and lived. (Ears would have evolved earlier, to capture such vibrations as the rattle of stones, the thud of feet, and the rumble of thunder.)

However vocal communication got rolling, it began so long ago that many vocal animals share a few universal "words" to this day. For instance, a low sound is still more menacing for most mammals, because it alludes to the speaker's great size. Furthermore, most mammals will fold those low noises into gravelly growls. On the other hand, if the same critters wish to telegraph an apology or a plea, they'll shift into higher tones to make themselves sound small, and smooth the texture into a whine. Many of us also employ the harsh screech of a distress call, commonly known as a scream. As a result, humans are able to eavesdrop on the aggressions and alarums of many other species—and presumably vice versa. The whimper of a young animal plays on a universal harp string; a snarl is a snarl is a snarl.

And speaking of harps, here's yet another theory on the origin of speech. This theory finds a foothold in my early babbling: That babbling was musical.

At six or eight months of age, I began to tap various parts of my mouth together, chopping the air into rhythmic bubbles: Da-da, ga-ga, ba-ba. Often, my hands would twitch to the same rhythm. I went musical. My mother appealed to my ardor for melody when she communicated with me. Whether she was reading me the grocery list or saying "Ga-ga, ba-ba" back to me, I was more inclined to shut up and listen if she delivered it in singsong style. Universally, human infants prefer to bestow their gaze on a mother who sings, as opposed to a talker. Also universally, human mothers regulate their infants' behavior with the tone and rhythm of their voices—with music, not with vowels, consonants, and verb tenses. To invigorate offspring, mothers run their voices up the scale. To pacify, they roll the notes gently downhill. To hold an infant's attention, they mix the two, in a roller coaster of sound. So, is it possible that humans are fundamentally hummers who diversified into talkers?

My own infantile brain did seem to mirror such an order of progress. I was born with a version of perfect pitch, as are all humans. Most of us lose it as our brain commits itself to spoken language. Does this imply that my own development retraced the development of language in our ancient human ancestors? Did evolving hominids derive speech from song? Music making is one of those behaviors that turn up in all human cultures. The world over, humans sing and drum, often adding physical twitches and jerks known as dance. And mothers the world over, even those who confess tone deafness, sing to their offspring.

Even as humans grow and learn words, we maintain this tonal version of communication. Most of us have had occasion to shout a warning to another human, but discovered that the mouth engaged quicker than the language. What came out was something like, "*Uh-uh-uh!*" Yet the signal transmits, based on tone alone. Similarly, studies have shown that both children and adults are highly literate in the emotional content of foreign speech. We may not understand a word of the lingo, but the emotion behind it is perfectly clear. And pair-bonded humans often exploit the efficiency of tonal communication to spare themselves the tedium of pronouncing actual syllables. I wouldn't want to rely on tone for arranging surgery, but it works admirably to request a repeat or to verify that my mate wants his coffee warmed up.

It wouldn't have been a big jump for early humans to add music to a base of simple calls. We are rhythmic creatures. Just walking out to gather roots would have been something of a symphonic experience: Feet slap the ground. The tide of breath rushes in, then rushes out. The arms swing. Humans in a group, ever social, will often adjust their pace subconsciously to synchronize their rhythms. And so before I could walk, before I could talk, I acquired music. The implication? I don't know. The foundation of language? Could be.

But why? Why prefer a babbling brook of sound to the occasional grunt, growl, or bark? What's the advantage to using the voice as a rhythmic instrument?

One tidy explanation finds its inspiration in the exceptionally feeble human infant. The human is born so helpless, and remains thus for so long, that a mother is occasionally forced to set it down. By contrast, a chimpanzee baby is born with the strength and instinct necessary to grab a handful of mother fur, and never let go. This leaves a mother's hands free to feed herself and perform life's other chores. But the human mother who needs both hands must set down her infant. Generally, human infants hate that. And so, the theory goes, humans evolved a pattern of vocalizing as a substitute for physical contact with a protesting infant.

This is nothing new, in the animal world. I recall the cows of my youth dropping a calf and immediately commencing a low-pitched grunt. Every breath brought forth this soft *moo.* The new calf absorbed the sound and would evermore know its mother by her voice. Mothers in the world of seals and sea lions do the same, with their pups answering in call-and-response fashion. Ring-tailed lemur mothers purr to their infants. So it's not far-fetched to suggest that humans might have stumbled upon the same method of remote control. When unable to soothe offspring directly, the human mother could produce a stream of syllables that would link her to the infant. With rhythm and changes in pitch, she could soothe, stimulate, and distract. And so a "musilanguage" was born.

Evidence favoring this scenario lingers in the human brain. There is that curious fact that newborn humans track changes of pitch in music, and in regular speech, too. And the fact that they're more interested in a singing mother than a talking one. Newborns are also able to distinguish one melody from another. Some research also shows that hospitalized infants fare much better if they're treated to music. And a distressed infant, even when he's being held, will settle sooner if he's rocked with the rhythmic sway of dance. These are hints that we evolved to rely upon that human-to-human link of music.

And if our ability to identify a tone weakens as we mature, other musical scraps seem to lodge in our brains for life. All humans are literate in the emotional content of music, even the music of other cultures. We all feel a physical response to music's rhythm, too. And a growing stack of experiments suggests a connection between music and the body's basic functions, including these preliminary findings: Soothing music reduces the amount of stress chemistry circulating in my blood. It strengthens my immune response and lowers my blood pressure. It improves sleep in elderly humans. A further batch of studies indicates that music also acts on us as social animals, causing us to feel connected. One of these found that humans who make music as a group experience a pleasant and painkilling endorphin flush. Experiments also show that humans are more eager to help one another after listening to soothing music. Therefore, as you might expect, many cultures use music to unite themselves for causes as diverse as healing, pair-bonding, god appeasing, hellos, goodbyes, and warfare. The rhythm and repetition inherent in music make it easy for a group to learn and to synchronize. (Spoken rituals are much harder to coordinate in a group. My own culture features only a few spoken rituals, such as the Pledge of Allegiance and various god appeasements and appeals. We've added strong rhythms to these but still make a hash when we try to speak them in unison. By contrast, our musical incantations, like "The Star-Spangled Banner," we deliver with one voice.)

If music did evolve as a means for adults to manipulate infants, it certainly has spread successfully to other areas of human life. It toils in the service of play, as my friend Monica and I sit with a glass of wine and a songbook, yipping the harmonies of country classics into the night. It serves as cheap social glue, as I mentioned, uniting us in preparation for football games, protest marches, labor strikes, birthday celebrations, religious rituals, and other ceremonies. It labors in support of mating, as I saw firsthand among my musician friends in younger days. In my culture, many human pairs adopt a piece of music as their own, one that symbolizes the early stages of their pair-bonding. And in some places, music has served as long-distance communication. Both yodeling and the alpenhorn, an extravagantly long wind instrument, are now fading from practical use in the Swiss Alps, after a venerable history of transmitting information between mountains. (Some languages that sound like music, however, are not. The world's many long-distance drum languages, as well as Silbo, a whistled language of the Canary Islands, aren't music but rather spoken languages encoded as drumbeats or whistles.)

Any one of these uses for music—mating, bonding, long-distance communication—constitutes a reasonable case for inspiring the world's first human music. Music would have served the cause of mating, one theory goes, by showcasing an individual's creativity and intelligence, both of which are promising features in a partner. Others have proposed that music evolved as a substitute for "social grooming." When troops of humans became too large for the members to bond by combing one another's fur, music emerged as a substitute, whereby you might bond with a dozen troop-mates simultaneously. The fact that chimps currently use drumming for long-haul communication argues that it might have served our ancestral hominids just as ably.

Wherever it came from, music came to stay. And some believe it was the parent of language itself. Did the hominids sing or hum or babble in rhythm before we could speak, just the way my own vocalizations progressed? It may be one of those proposals we'll never hear answered.

But also for what it's worth, humans are surprisingly attentive to the musical element of spoken language. Two humans from different cultures can often extract the emotional gist of each other's speech, based solely on the "melody" of the delivery—just as we can with music. And in "tonal languages," which include Mandarin, Punjabi, and Navajo, it's the musical delivery of a word that determines the meaning. In Mandarin, most famously, "Ma ma ma ma" can mean "Did mother scold the horse?" But only if you deliver it with the correct melody. It could also mean "Horse horse horse horse." Incidentally, Mandarin speakers appear to hang on to their childhood pitch perception better than speakers of nontonal languages like mine, as though exercising the pitch muscle keeps it strong.

THE FULLY VERBAL HUMAN

After an infancy of vocalizing, a childhood of verbalizing, and years spent refining my physical signals, I arrived at adulthood. I am a verbose and musical representative of my species—and my gender. As a human female, I wield my communication tools a bit differently than do males.

Actually, contrary to popular rhetoric, the distinctions aren't as interplanetary as they're reputed to be. In a pop-psychology book, *The Female Brain*, I read that I speak 20,000 words in a typical day while my mate barely gets out 7,000. Similar claims (with dissimilar numbers) proliferate in self-help books and on the Internet. But when I tried to find scientific references for this field of study, I was underwhelmed. In fact, the latest investigation, based on automated recorders worn on the human body, found no statistical difference at all between the word count of males and females. Individual males did, however, claim the titles of most taciturn (about 500 words a day) and most verbose (47,000). But on average both sexes put forth about 16,000 words a day.

There is evidence for two real differences, however. The average female matures with a small advantage in the ability to reel up words from memory. And a clear split divides what males and females talk about. In this case, the trend from a few hundred studies is that females broadcast more information about themselves and other humans than do males. Male communication more often concerns objects.

If the sexes communicate differently, at least they speak the same language. This isn't the case for all animals. I'm charmed to find that among a few primates, each sex has its own lexicon. If a male and female diana monkey spot a leopard at the same moment, they'll issue his 'n' hers alarm calls. If the female hears the male's warning from a distance, she'll repeat it, but in female-ese. How—and why—these parallel systems evolve is a mystery. Most birds are segregated singers, too: Only the males produce song, those complicated tweedles that are meant to attract females and intimidate other males. News to me is that some birds segregate their more routine calls, as well. When a male issues the foraging call, which informs his mate of his location, he might say the equivalent of, "I'm over here." Rather than respond in kind, she'll say, "Here I am."

In spite of these precedents, I'm still wowed to discover that a handful of human tribes around the world also maintain separate male and female dialects of their language. Egad! I first encountered sex-specific language in a description of Lengua Indians in Paraguay, but I put it aside out of fear that the missionary reporting the situation might have been suffering from the fevers or the vapors or something. Picking up the subject later, I found this schism has developed more than once. The males and females of the Yanyuwa of northern Australia share the roots of their words, but then tack on all manner of prefixes and suffixes particular to each sex. The Yana Indians of California flip around whole chunks of verbiage, depending on their sex: From the mouth of a female, a grizzly bear is a *t'et*. But a male would call the same animal a *t'en'na*. The Yana females speak the female dialect to one another and to males; males speak female-ese to females and male-ese to other males. Given the obstacles already confronting the human pair-bond—the differing brains, the perishable love chemistry, the conflicting reproductive agendas—it seems an unnecessary cruelty to ask the two sexes also to speak different tongues. Really, isn't it hard enough already?

SENDING DECEPTIVE SIGNALS

I remember my first conscious effort at deception with ghastly clarity, probably because it was such a failure. I had appropriated one of my brother's embossed pencils, whittled his name off it with a knife, and declared it my own. Was I ever dismayed to discover that declaring something to be true doesn't make it so. "It's mine. I didn't carve his name off it," I wailed doggedly, believing this is how lying was conducted but sensing that something had gone terribly wrong. I was offering up the right words, but nobody was buying them. I had so much to learn.

Young humans are graceless liars. Shortly before my pencil episode, the brother in question had sworn a solemn oath that he had received on Christmas a peppermint stick the size of a fence post. Never mind that we had all spent Christmas together and that no one else recalled a one-hundred-pound candy cane. He would not waver.

Lying takes years of practice to perfect. But it's a worthwhile endeavor. Deception is not a subset of communication. Recall that communication is about manipulating others to your own benefit. So I propose that in the world's first conversation some animal mother called her infant toward a mound bustling with nutritious ants, and in the second, she told the mother next door that the ants were rancid.

When two animals covet the same ant, fruit, or nest hole, it behooves them to use communication (rather than tooth and claw) to deflect each other from it. Thus, many animals are adept fibbers. The male barn swallow who discovers his female in the embrace of another male will screech out, "Predator coming!" The cheating hearts will fly apart and take cover. The Formosan squirrel from Taiwan takes a proactive approach with his lies, shouting "Predator!" after his own mating bout. This sends competing males into the trees, delays the female's next mating, and gives the liar's sperm a head start. The burrowing owl of the American West, when it hears a badger approach its den, issues a call that mimics the buzz of a rattlesnake. Sometimes an animal's entire life can become a lie. Male orangutans who aren't able to win their own territory become homeless wanderers. In this case, their bodies stay small and slim—female looking. These cross-dressers slip past bullying males and sneak up on unsuspecting females. The result of this physical lie is an impressive reproductive rate.

All of these are examples of instinctual or evolved lies. But occasionally an animal (besides the human) seems to make a conscious effort to mislead another. Among baboons, it's the young who seem most devious. One little devil reportedly learned to deflect his mother's wrath by standing erect, eyeing the horizon with terror, and screeching an alarm call. Another youngster specialized in false accusations of child abuse. This prodigy would watch a female baboon dig up a juicy root, then screech, "She hit me!" His mother, fooled into a protective rage, would barrel over and chase the "abuser" away from a hard-earned meal.

The chimpanzees are expert liars, too. One report involves a lothario who was leaf clipping toward a fertile female and then stuffed the leaf in his mouth when the boss chimp happened by. "I wasn't leaf clipping!" Chimp scholar Frans de Waal reports on a low-ranking male who was displaying his penis to a female when a high-ranking male came on the scene. The would-be-suitor clapped his hands over his crotch, blotting out the message. De Waal also witnessed a male with ambitions for higher office trying to disguise his physical communication. When the chimp spotted the alpha male, instinct spread a "fear grin" across his face. But this self-aware fellow, consciously hoping to conceal his mental state, reached up and squeezed his lips shut over his teeth: "I ain't scared of you!" Also, chimps of both sexes observe uncharacteristic silence when copulating within earshot of the alpha male.

We humans also do a lot of our deceiving to hide copulations. After all, if a female wants to copulate with a male who is already pair-bonded, why try to fight the first female off when you can just duck behind a shrub with the male? If the other female should appear, you can always squeeze your lips into an expression of innocence. Of course, humans lie to protect other important resources, too. In my culture, no employee in his right mind informs the boss human when he starts looking for a better job. He conceals his personal goals from the boss, who might punish him for disloyalty. Even when I'm hunting and gathering in the aisles of Macy's I endeavor to protect the best resources for myself. I do not shout out, "Hey, ladies! Big rack of Liz Claiborne on sale here! Come on over!"

This is all just the tip of the iceberg of lies. Humans lie all day long. We do it so often it doesn't even require much effort: "I'm fine, thanks." "I don't mind waiting." "What a cute baby." "Iraq has weapons of mass destruction." A recent experiment monitored pairs of strangers killing time in a waiting room, and found that, in a human, the lies flow like water. In the ten-minute test period, 60 percent of the subjects rattled off an average of three fibs apiece. As you might expect, humans with differing personalities, or differing twenty-third chromosomes, dispense their deceptions differently. Extroverted humans are handier with a lie than are introverts. And while males and females seem equally prolific, the sexes tend to dole out deception for different reasons: Females more often lie in the interest of maintaining social harmony and soothing others; males are more likely to dissemble in a way that brightens their own image.

Researchers have found that spitting out the words is just half the battle in unfurling the successful falsehood. The body is harder to manipulate into a misleading posture. Consider the "Duchenne smile." Named for a scientist who mapped facial muscles, the Duchenne smile is an involuntary expression that involves the mouth, cheeks, and eyes. It turns up when we're genuinely happy. And it's extremely difficult to fake. Most humans can manage the mouth and cheeks, but even professional models have a hard time forcibly tensing the muscles around the eyes without looking like they're about to hit someone. And forget about holding that look. An honest and natural expression can live for about four seconds on the face before it starts to quiver and crumble. Add to that the tension that creeps into a self-conscious voice, and the other twitches and tells that flicker through our faces, hands, and feet, and lying can become a rather strenuous exercise. But from childhood on, we practice doggedly. And we do improve. I can't say that I, personally, being both an introvert and a female, would perform any more convincingly if I stole a pencil today than I did as a child. But I think I've learned to whip out a "No, that didn't hurt," and an "I had a lovely time," at a reassuring tone and pace.

For better or worse, humans are dismal lie detectors. Even our best scientific tools—which measure everything from a teensy change in temperature around the eye, to brain activity patterns, to "micro-expressions" flashing across a fibber's face—even the best miss at least one lie in ten. And human practitioners are far worse at lie catching, judging from controlled experiments. The public, most police, and even judges only catch a lie about 50 percent of the time, which is the same rate they'd achieve predicting the outcome of a coin toss.

So lying does, in fact, work. Crimes of communication do pay. And although other apes and animals can spin a yarn now and then, no creature on Earth can match the eighteen-lies-per-hour rate demonstrated by the animal we might call the Unreliable Ape. Humans communicate at a breakneck pace, and pell-mell prevaricating is part of the package.

SENDING SIGNALS BETWEEN SPECIES

We tend to see each animal species as an island, communicating only within its own membership. But the truth is we're all monitoring the waves of communication that spread from one island to the next. We collect information wherever we find it and adjust our behavior accordingly. The squirrels in my yard, for instance, rear up and survey the landscape whenever the resident crows utter an alarm call.

What threatens a crow could, after all, threaten a squirrel. Squirrels who act on the free warning live longer and prosper.

I suspect that many animals are sensitive to the basic calls and body postures of other species. Perhaps because I grew up on a farm, the body language of mammals seems as obvious to me as if they spoke English. The aggressive posturing of a rat has much in common with that of a hyena or a bear: The head comes down, the neck fur goes up, the ears rotate forward, and the eyes bore into the opponent. Because so many mammals share similar physical "words," my species is probably not the only one who can read the body language of other creatures. My dog is perfectly attuned to the glare of a cornered woodchuck and the chatter of its teeth, and replies with frustrated barking.

Sometimes, the signals between two species become so rich and rewarding that the animals formalize an arrangement. Animals with a symbiotic relationship have to develop messages that each side understands, if the treaty is going to last. A cleaner wrasse on a coral reef first advertises his services with bright blue and yellow coloration. He'll add dance steps if trade is slow. A parrot fish who wishes to be cleaned approaches more slowly than she might if she were feeling predatory. She hovers in front of the wrasse's territory, opens her gills wide, and stretches her fins. As the wrasse scours his client, he may even polish her dentistry, but he vibrates his own fins to remind her he's not to be eaten. Both species attend to their communication, and both prosper. Something similar must transpire between the Nile crocodile and the dainty plover, as the bird picks wildebeest gristle from the teeth of the reptile.

And rarely, preciously, a similar agreement grows up between humans and another animal. The honeyguide of Africa is a nondescript bird with a knack for spotting bees' nests. Unable to breach the bee defenses herself, she instead flies in search of assistance. When the bird finds a honey badger, she issues a distinctive squawk and commences a swooping series of fluttery flights back toward the hive. The badger, who resembles an overgrown skunk, follows. He subdues the bees with an anal secretion (Zounds!), and demolishes the hive. The honeyguide watches from above, descending to clean up the remnants when the badger is sated. How do humans come into it? The honeyguides have also evolved to seek out Pygmies, who are happy to stand in for the badger.

More complex was the dialogue that developed between orcas and one family of humans in Eden, Australia, during the whaling era. Three local pods of orcas would cooperate to herd baleen whales toward Eden's harbor. Then a few orcas would sprint toward a farm owned by the Davidson family and roust the whalers by slapping their tails on the water. (The Davidsons employed native Yuins as a whaling crew and had adopted their brotherly sentiment toward the orcas. This probably allowed trust to germinate between the two species.) The whalers would jump into their boats and row behind the orcas to reach the captive whales. Upon killing a whale, the Davidson crew would grant the orcas time to eat its lips and tongue. In the event that the Davidsons killed a whale without assistance from the orcas, it was then their turn to slap their oars on the water, signaling that tongue and lips were served. This interanimal agreement endured for a century, stuttering to an end as foreign whalers arrived, began shooting the orcas, and caught so many whales that the entire predator–prey relationship collapsed.

Perhaps most impressive is the conversation that has lasted fourteen thousand years between humans and dogs. Many domesticated animals are too stupid to be very interesting. But dogs, despite their brains shrinking 10 percent during their descent from wolves, are social animals with an inbred desire to swap information with humans. One of the resulting differences between wolves and dogs is that dogs understand the human pointing gesture. Point out a biscuit for a wolf, and it will stare blankly at you. Point for a puppy, and it will commence snuffling for the subject of your message. And the communication flows both ways. Probably the dog's first message to humans was, "Someone's coming!" For many dog owners, this remains the most valuable thing a dog has to say. But eons of breeding have left the species able to convey a great deal more. The animal has evolved in step with human culture. It can now function as human hands, opening doors and dialing 911, or as eyes, guiding blind humans through traffic and revolving doors.

Humans can communicate with innumerable other species in a limited way. Solitary dolphins sometimes grow fond of local humans and develop a repertoire of attention-seeking signals like those of a dog or cat. They bump swimmers or tug their snorkeling fins. Manatees, those cumbersome sea cows of the tropics, are happy to commune with humans if the humans scratch their hides. The crows whose territory encompasses my yard understand me perfectly when I call "Crow-crow!" Within seconds they arrive in the oak tree to see what goodies I've left on the shed roof. And when they veer close past my office window, I understand them perfectly: "Goodies, lady. Bring us goodies."

Because communication indicates an unmet need, why is it that I heed the demands of crows or agree to scratch the back of a manatee? For humans, with our social drive and our huge brains, bonding is a need. And apparently we're not too particular about the species. So when my crows make the effort to signal to me, I am deeply flattered. I am putty in their talons. And I hasten to meet their demands.

Needs. It's back to needs. As a fetus, I needed to grow, and so I cried out chemically for nutrients. Once I emerged into air, my needs grew in number—for food still, but also for warmth, for gas relief, for dry skin. My communication channels grew, too: I could flap and squawk now like a hungry bird.

As my brain grew, it wired up the controls for my tongue and lips. I began to whack my breath into cute little segments, *da*, *da*, *da*, my hands whacking out a rhythmic vocabulary of their own. And then I linked sounds to objects. Now I could call for my needs by name. My brain, and its burgeoning ability to convey my thoughts, was reenacting the ancient evolution of hominid communication.

Soon I was linking together my noises, following a grammatical pattern all humans seem to carry in our DNA. Now I was leaving behind most of my animal kin, heading into the rarified realm of language . . . or not. Perhaps I was joining a host of creatures, from prairie dogs to vervet monkeys, whose linguistic aptitude we've been too self-absorbed to appreciate.

It is a measure of how well we've met our needs as a species that we can now turn our attention to something as arcane as prairie-dog talk. But it also illuminates a need that is uniquely human—as far as we can figure: We feel the need to study prairie-dog language because it can inform us about our own origins, the social and ecological milieu that drove us first to the pebbles of speech, then on to the masonry and castellations of language. We feel the need to understand ourselves. If we didn't, we might, at this stage of our career as a species, sit on the porch in a rocking chair. With the status quo satisfactory, we'd have nothing to discuss. *Homo sapiens* would finally fall silent.

10
TOUGH AS A BOILED OWL:
PREDATORS

The human was once an important prey animal for such large carnivores as leopards, lions, and other cats; some bears; reptiles, especially crocodiles; hyenas; and a great number of parasites and microorganisms. Although *Homo sapiens* is a relatively slow breeder (and hence rare through most of its history), and not naturally a fat animal, carnivores nonetheless find the primate an attractive target due to its physical weakness and a large, fat-rich brain. The animal's social nature also makes it appealing to microorganisms that spread through casual contact, body fluids, or bodily waste. Especially since formalized agriculture brought humans together in large troops, these micropredators have been an important limiter on the human population.

But in recent centuries this animal's intelligence and tool use have reordered its position in the food web. Throughout most of its range, the human has reduced large carnivores to a shadow of their former populations. The human not only competes more effectively for prey animals but also stalks the competing carnivores directly.

In the past 150 years *Homo sapiens* has also targeted its micropredators for elimination. These still handicap the human substantially, killing millions each year (mainly young, old, and ill individuals). But both diarrheal and respiratory microbes are also on the wane, as medical tools reach the poorer human territories.

The human has not cleared the field, by any means. Most interesting is the continuous evolution of new microbial predators, whose brief generation times allow them to mutate new weapons much more quickly than their human prey can evolve defenses. HIV and SARS are recent examples of this dynamic.

However, an ever-growing population argues that while this primate may occasionally lose a battle to a predator, it's winning the war. The repercussions of this victory aren't fully known. One result is an unforeseen sense of lonesomeness. The primate is now making the conscious choice to let some of its large predators survive.

HOMO SAPIENS AS CAT CHOW

I expected that the breath of a polar bear would smell of something—old berries, fermenting seal, fermenting human. With his black nose two feet from my own, I smelled nothing but the clean Arctic air. I had been deployed to the frozen tundra to write about polar bears for a travel magazine. I was surprised to find that I, too, was the subject of investigation. The nose twitched. I probably stank a whole lot more than he did.

We *Homo sapiens* have killed off almost all our large predators. For the most part, we flee from looming monsters with shining fangs only in our dreams. But once in a while, the real thing appears. For a human raised in the sanitized woods of North America this can produce quite a shock. I met my predator on the western shore of Hudson Bay, Canada, where polar bears gather each autumn. There they wait for ice to form so they can go seal hunting. And there human tourists, safe in giant school buses on ten-foot tires, gather to watch them. It was on the back porch of one of these tundra buggies, searching the brown eyes of a bear, that I finally faced a predator who wanted to pull me apart and swallow the soft pieces. And realized it was nothing personal.

I looked down at the bear, who rose to place his paws against my porch wall. The hockey puck of his nose twitched in the stream of odor pouring off me. I could have reached over the boards and touched his claws. A few weeks before, a photographer standing here had hung a forearm over to shoot along the length of the bus, and a bear had come from behind and torn away that low-hanging fruit. I searched the eyes for a doglike warmth or a spark of curiosity. I found nothing at all. A brown eye looking at, well, a prey animal. Still motionless, I whispered, "Good bear. Handsome bear. Hungry bear." The nose sampled my breath, the brown eye betrayed nothing. I didn't move until he heaved his paws free of the porch, dropped back to all fours, and shambled away.

Despite having cleansed the planet of most large predators, humans can still do a decent job of behaving like prey. That freezing response is probably built into my genes. It happens when I'm walking alone in the woods and a stick nearby goes, "Snnnap . . ." My ears whip my head around to engage my eyes in the search, and I stand stock-still. Adrenaline swirls into my blood to fuel a superhuman flurry of action, if I should need to flee or fight. My pupils dilate to snare every photon's worth of information. This is the textbook response of a prey animal. When I surprise a white-tailed deer in the woods, her reaction is the same: freeze, and gather information. And likewise the mother grouse with chicks, and the rabbit, and the toad. Most predators excel at spotting motion and aren't so hot with detecting a pattern of color and shape. So if it's too late for a prey animal to creep away unnoticed, freezing is the best option.

Humans honed the freeze response during a long history of serving as cat food, wolf food, and even bird food. The entire mammal lineage, in fact, spent some formative eons skittering under the noses of predatory dinosaurs and reptiles. Size conveys some privileges, so as our ancestral species developed into larger primates, then into hefty apes, the number of predators chasing them declined. But even chimpanzees of the modern world have to sit with their backs to the wall. Leopards and lions lie in wait to ambush them; hyenas and wild dogs tear them limb from limb; crocodiles bide their time in rivers. Out of every hundred chimps, five or six will be caught and eaten every year. (And recently, humans have begun preying so heavily on chimps, which have become a status meal for some urban Africans, that the species may be destined for extinction.) Even the gorilla, at three hundred to six hundred pounds, isn't big enough to escape death by digestion, as leopards demonstrate periodically.

Our situation didn't improve much even after our ancient ancestors split from the other apes. The early australopiths like Lucy and Tuang Child were lithe, little bites, armed at best with stones and at worst with their bare hands, feet, and teeth. Their running speed was laughable compared to a lion's. Presumably, they fared poorly against large carnivores. Fossils are rare enough, let alone fossil hominids with lion teeth lodged in them. But according to a marvelous book called *Man the Hunted*, one bone bed in South Africa dating to Lucy's era holds the remains of predacious hyenas, leopards, and saber-toothed cats, plus the remains of their prey: baboons and pieces of Lucy's kinfolk. At another South African site, scientists have found the skullcap of a Lucy-like child bearing two puncture holes. The canine teeth from the lower jaw of a leopard fit nicely. And the empty eyes of Tuang Child tell another sad story: This three-year-old 'pith was apparently snatched up by a large eagle and carried away for dismemberment. Beak punctures in the eye sockets match those that eagles make when they scoop eyes out of modern monkeys.

■ ■ ■

Fortunately our ancestors did evolve larger frames. Predation can do that to you. One of the greatest forces for change in the DNA of any animal is predation pressure. If leopards catch all the slow hominids, then hominids very quickly evolve into faster animals. (Of course, this causes starvation of the slowest leopards, so that species also becomes faster. This "arms race" dynamic drives each animal's evolution in whatever direction produces the greatest number of healthy offspring.) So, predation may have pushed hominids to evolve bigger bodies because the biggest individuals were the least likely to be attacked. Or maybe predation selected for hominids who were inclined to throw rocks, leading to a tool-rich future. Or perhaps stealthy leopards weeded out the dull-witted ancestors while the biggest-brained ancestors escaped to breed. At any rate, hominids grew, and probably shook off a few predators in the process. Eagles, for example, massacre small monkeys in Africa, but find the prospect of butchering an ape more daunting—not impossible, but certainly no picnic. Eventually, hominids, with their crafty brains and their clever tools, rose to join the ranks of nature's elite predators.

My reproductive capacity is evidence of my predatory status. Those creatures who are accustomed to being eaten give birth to big families. Rabbits breed like rabbits, as do many small mammals. They must, in order to keep up with attrition. Even the speedy ungulates—gazelles, wildebeest, deer—drop a kid once a year to combat the constant pressure of carnivores. But humans, when we live without infant formula and agriculture, seem to breed on a four- or five-year plan. That's very slow and a very loud argument for our high rank on the predating scale. We prey upon far more than we are preyed upon.

That doesn't mean we're inviolable. An adult skull from the Dmanisi site in eastern Europe demonstrates that carnivores caused headaches even for stalwart *H. erectus*, the world-traveling *Homo* of about 1.8 million years ago. On this skull, the puncture holes make a lovely scabbard for a set of saber-toothed tiger teeth. In China, another pile of *H. erectus* skulls appears to be the discards of a protracted feast enjoyed by hyenas, who have a signature method for cracking bones. The skulls were gnawed open for their oily brains. (Predators must have loved the steady enlargement of the hominid cranium. Brains are a prize in any prey, and hominid heads were positively bursting with them.)

Even after we *Homo sapiens* upgraded our tool kit and lit a fire between ourselves and our nocturnal enemies, we were still feeding the cats—and wolves, hyenas, and so on. It wasn't until we invented the cheap bullet that some humans could begin to relax in the great outdoors. Finally, in the past few centuries, those cultures with access to lead ammo could plug away at their enemies without the worry that they'd be spending the next six months making replacements for all the lost arrowheads. In short order, humans cleansed great swaths of Europe and North America: The few surviving wolves, bears, and big cats retreated to patches of wilderness. We humans prowled our territories with impunity.

Skulking in mountain refuges and dismal swamps were the remnants of our predators. A few cougars continued to eat deer in the North American West. Wolves chase down moose in the dark forests of Canada, Scandinavia, and Russia. The big bears, too, have contracted into the northern realms where humans are scarce. Only in the less-humanized parts of Africa, India, and Asia do large carnivores like leopards and tigers continue their pursuit of human prey.

Well, they also haunt the wealthy suburbs of North America. As guns vanished from the suburbs, a few curious carnivores poked their noses back in. For the first time in a century, the human animal in North America now has to watch its back. In New York State, black bears first raided bird feeders, then one plucked a human infant from its carriage in 2002. In California, cougars took down seven full-grown humans who ventured into the hills to hike or bike in the early 1990s.

But as their predators rebound, the humans are doing a most un-animally thing. In the richest cultures, where bullets are cheapest, humans are waxing nostalgic for their old tormentors. In half the western states where cougars are accumulating, humans have voted to protect them from all hunting or from hunting with dogs. In 2002

The New York Times ran a story that stopped me in my own tracks: In Colorado, where homes are pushing deep into cat territory, cougars have been eating pets—*and the humans aren't complaining*. "We moved into their territory," was the feeling expressed in the article. "Yes, one ate our cat. But it wouldn't be fair to hurt them."

Whoa. Let's review for a moment: Would a troop of chimpanzees move into the territory of their sworn enemy and then decline to kill them off when given a free shot? No. Would orcas? Um, no. How about wolves? Unlikely. No, it's only our species that can pin a dangerous predator to the ground, then engage a moral muscle and release the enemy. Admittedly, it's not a common behavior for humans. It arises mainly in wealthy cultures where humans have had the luxury of studying ecological patterns beyond those that feed and shelter us. In my rural state, humans are divided into two warring camps over such issues. One camp sees the advance of coyotes as a triumphant return of nature and an end to centuries of human-centered policy. The other sees the coyote as a threat to their cocker spaniels and white-tailed deer. In other parts of the planet, the same division attends the return of the wolf or the increasing appearance of cougars in the suburbs. Many humans see no reason to share their habitat. But that any humans at all would take the side of a competing predator is something of a marvel.

To some degree, the human animal is willing to take a risk—an unnecessary risk—of becoming cat chow. Strange beast.

A ROGUE'S GALLERY OF CARNIVORES

So large carnivores remain among us—sometimes because we're unable to track them down and sometimes because we've decided to tolerate them. Thus I have stood nose to nose with a polar bear, willingly. In the Florida Everglades I have crouched within striking distance of a ten-foot alligator, watching for the faint motion of its ribs inflating with breath. (I was new to Florida and its ubiquitous reptiles. I had walked so close because I presumed the animal to be a plastic gag.) In a Mongolian desert I have stood breathless as a Gobi wolf trotted out of a wash and crossed the yellow sand in front of me, casting an appraising glance as he passed. A normal animal would not treasure, let alone seek, such encounters. I find that crossing paths with a predator adds depth and context to my human existence. I have counted these meetings among the shining moments of my life.

A word about the word: A "predator," in biological terms, is an animal who kills another to eat it. Although I kill both cows and cockroaches, I am a predator only of cows. I, in turn, might be killed either by a whitetip shark or bad sushi, but only one of those qualifies as a predator. Animals (and microbes) in two additional categories can do me in—parasites and competitors—but we'll deal with them later. Just now, I'm interested in those organisms who wish to digest me.

The human's peak predator is probably the crocodile. The exact number of humans that crocs eat each year is hard to pin down, because they tend to hunt in undeveloped areas where humans don't keep detailed records. Human prey is a good choice for crocs. Our species is big enough to make a hearty meal but free of sharp horns, claws, and teeth, and small enough to subdue quickly. As with the other midsize animals they catch, crocs reduce human prey to bite-size pieces by biting onto a limb and rolling their entire bodies in the water until the twisted part tears free. A whole flock of crocs can feed on a single kill, with each animal latching onto a different piece and rolling until the prey disintegrates. It's estimated they eat thousands of humans a year.

Second on the list might be the tiger. And that's almost entirely due to the tigers of Sundarbans, a forest on the Bay of Bengal in Bangladesh. In these tangled mangrove swamps fish haunt the murky water. Humans in wooden canoes come to prey on the fish. And tigers, padding across the roots and hidden by the glossy leaves, come to prey on the humans. The biggest male Bengal tigers can weigh seven hundred or eight hundred pounds. Even the smallest female presents two hundred pounds of muscle and bone, tooth and claw. To knock down prey, a tiger hurls his full weight like a cannonball, then locks a kill bite onto the nape or the throat. Usually tigers stalk more challenging game than humans. But the Sundarbans tigers have become human specialists. They take some three hundred a year. Before the introduction of bullets to the region, their harvest was estimated to be as high as fifteen hundred a year.

The leopard, whose range encompasses southern Asia and tropical Africa, is one-third the size of a tiger but nearly as lethal. Again, statistics are elusive, but a survey of one state in India found that the spotted cats there were killing about thirty-four humans a year. Another Indian state sees about fifty attacks a year, not all of them fatal. With fine night vision, these cats rely on stealth, not size. Stories abound of leopards slipping into houses to carry away slumbering humans and even dogs. Like most cats, they kill with a sharp bite to the neck, or a throat-hold. Leopards prefer to store uneaten meat in a tree out of reach of hyenas, so you might expect them to target small prey. The truth is this predator is so furiously strong that hauling a mature human into a tree is no problem.

The other big cats also catch an occasional human. In Africa, individual lions sometimes develop a preference for our species. But more often, they seem to turn to humans when humans have eliminated too many of the hoofed animals lions normally eat. For instance, as the population of *Homo sapiens* in Tanzania exploded in recent decades, the number of zebra, hartebeest, and impala tumbled. Concomitantly the number of humans tackled by lions has soared to about one hundred a year. (About two-thirds of those attacked die.) Big cats of the Americas, the cougar and jaguar, occasionally hunt humans, though their success rate is probably lower. (Data on jaguar attacks are especially slim.)

In addition to cats, a handful of other carnivorous mammals continue to nip at the fringes of the human population. In Europe, wolves still take human young from time to time. Once in a while hyenas hunt humans in Africa. Black bears in North America attack humans for prey. (Brown bears more often attack humans in order to defend their food or a cub—but once the intruder is dead, why waste good meat?) Polar bears will also take advantage of a stray human.

The crocodile isn't the only reptile to view humans as prey. The alligator is another, and so is the Komodo dragon. Each can kill and eat humans of any size, but compared to crocodiles neither puts much of a dent in the human horde. And then there are the fishes, sharks to be specific. Most sharks don't consider humans especially palatable. They might bite one if they mistake it for a flailing fish or seal, but they don't seek humans out. A few species, however—oceanic whitetip, great white, tiger, and bull—are happy to harvest humans. I have seen the whitetips in action, and it's a haunting phenomenon. I was reporting from a research boat parked for two weeks in the emptiness of the eastern Pacific, where scientists studied deep-sea vents. Every day a few humans climbed into the *Alvin* submarine to be lowered into the water. This operation required two or three additional humans to dive underwater and unhook a cable. Then they'd surface and throw themselves into a waiting rubber dinghy. We dumped no garbage and shed no blood, but the sharks found us. First it was one whitetip, a slim blue streak circling the ship. Then two more. And then seven. Each time the *Alvin* was lowered, the sharks would slip through the waves like shadows, waiting for the humans to enter the water. The rest of us stood breathless on the deck, straining our eyes for the flash of blue so we could shout to the swimmers. But as relentless as these predators are, fairly solid records suggest that sharks attack only sixty or seventy humans a year, worldwide.

As for birds of prey, it may be that humans have finally outgrown their raptor predators. Rarely, an African crowned eagle will attempt to snatch a small child, as birds apparently have done for millions of years. Within the past few decades researchers have found a fragment of human infant skull in a crowned eagle nest, and have reported on a seven-year-old child injured by one of the birds in Zambia. The crowned eagle tops out at twelve pounds but can kill and carry prey many times heavier. They kill with their feet, which are preternaturally strong.

This, too, I have seen firsthand. When I was a kid, our pet screech owl Wowl was seven inches tall and weighed about six ounces. He was mostly feathers. But when he landed on your bare shoulder, you knew it. His feet were outsized, and the talons more so. His grip pierced like a dental tool. Periodically, Wowl would turn predatory on the humans he loved. He would still coo in our ears and preen our hair with his hooked beak. But at night, when only my parents roamed the house, he would wing silently down a hall and momentarily set those claws in the back of their heads. As much as this was surprising, it was also painful, and when he was in those moods my parents took to wearing headbands with eyes painted on the back. Fooled, Wowl desisted.

And that owlish digression turns us to the subject of defense. The best defense is a good offense, of course, and humans have excelled at offending our predators into oblivion. But when a rare survivor comes at you, what is a human animal to do? Well, as we do with so many other challenges, the human under attack generally reaches for a tool. In fact, our defensive use of tools raises an interesting question: Did the early hominids take so maniacally to tools because, under-armed as they were, tools improved their odds of beating off the predators?

I recently encountered an insightful little piece of work at a conference. In the back of a room full of research summarized on large posters stood a psychologist from Pennsylvania State University. His poster was deceptively simple but profound: He had combed the globe for stories of "wild beast attacks man." Studying 173 incidents, he concluded that humans with no tool in hand died two-thirds of the time, but those with a defensive device of any sort died only half the time. That's a big difference in survival rates. It's the kind of difference that can shove evolution rapidly in a new direction—an armed direction. For what it's worth chimpanzees also use tools against snakes and other frightful animals.

Tools aren't a human's sole recourse against predators, of course. We also have access to the standard mammalian package. This includes adrenaline, which, when released by a surprise, underwrites an emergency expenditure of energy by the muscles. We can spend this energy to run, climb, or fight. If it's too late for these reactions, the human will sometimes freeze, hoping that the motion-sensor eyes of the predator will count us as just another tree, another rock. Should that fail, we may resort to threats, feinting and waving our arms to look large—just the way a nervous chimp will inflate his outline by erecting his fur, or the humble fruit fly will raise her wings. And like the rabbit, whose options are exhausted by the time she's in the fox's jaws, we also scream. Sometimes, that's sufficiently startling to the predator that he'll drop his prey.

But again, this is not a daily concern for most modern humans. We have purged our territories so systematically of large predators that most humans never face an animal that intends to eat them. That's a rare situation for an animal. Most critters spend their lives looking over their shoulders. They are almost always at risk. The handful of carnivores in our privileged position are called superpredators, or apex predators. Tigers are one. Great white sharks are another. And so are we. But we weren't born that way. We scrambled to the peak only recently, on the ladder of our tools. Should we consider this—our superpredator status—the natural state of the human? We came by our tool use naturally, after all, as naturally as does any crow or chimpanzee. Therefore, is our use of tools to vanquish our predators natural? If not, at what point did our defenses cross the line and become something supernatural? Perhaps it's immaterial. More interesting, the Compassionate Ape has the capacity to empathize with other species and to spare their lives. It's this moral capacity that looks so out of place in nature.

A LESS ROGUEISH GALLERY OF MERE COMPETITORS

Before we turn to our microscopic and most fearsome predators, I have to give our competitors their due. Scientists divide interspecies relationships into categories. Predators eat me. Symbionts dwell in harmony with me. And competitors are those species who covet the same resources I want. This competition between species can become violent, even fatal. Thus, while the common cobra is not a predator of humans—it has no desire to eat us—it is nonetheless equipped to kill a human who interferes with its life, liberty, and pursuit of happiness.

Our competitors are more lethal to us than our animal predators, at this point in human history. Snakes, for instance, bite half a million humans each year, killing 125,000 of us—far more than the crocodile. And this is probably a gross underestimate, given that snakebites are most common in the poorest parts of Africa and Asia. Humans bitten in small villages are unlikely to visit a hospital, from which their plight might be reported to the World Health Organization. All these deadly snakes aren't out to devour humans. It's just that snake territory and human territory often overlap. Each species wants to feel safe in this territory and each perceives the other to be a menace. Snakes most often face off with young male humans, because both parties spend a lot of their time in agricultural fields. The winner is the species that strikes quickest in any given contest.

Even more deadly competitors than snakes are the hymenopteran insects: the ants, bees, and wasps. These animals sting when competing animals menace their hives or their lives. And while most humans bat away the insect and rub the sting, a few percent of stung humans break out in an allergic reaction. This can culminate in suffocation as the throat swells shut. And this is no predator. Just a determined competitor.

Lagging well behind the hymenopterans and the snakes are an assortment of competitors whose sensational weapons win them outsized reputations: Scorpions take a few hundred humans each year; spiders make a token showing. The piranha, sharp-fanged fish of the Amazon, is in reality a weak competitor. Although a piranha won't say no to an easy meal from a human corpse, it bites live humans mainly to drive them away from its brood.

Perhaps the dogs that kill humans fall into the "competitor" category, as well. They're fairly perilous to humans, tearing the skin of nearly a million each year in the United States alone. Dogs almost never eat their human victims, so they don't see us as prey. Since certain breeds are more inclined than others to injure or kill humans, it may be that some biological drive—perhaps competition over territory—is at work.

A MALIGNANCY OF MICROBES

It's the growling animals with glowing eyes who populate the dark-night terrors of a human. It's the snakes and spiders who inspire an otherwise rational primate to leap skyward with a screech. But in reality, our most dangerous predators are hairless, scaleless, toothless, and teensy.

Living as I do in a tool-rich culture, I've always succeeded in fighting off the microbes who attempt to devour me. But they've tried. Many have tried. I've endured cold viruses who opened the door to their friends, welcoming bacterial invaders into my lungs. One winter I wheezed and gurgled for days before someone convinced me I had pneumonia and had best see a doctor. I've had simple cuts that became staging grounds for strains of bacteria capable of poisoning my blood and dropping me like a sack of wet sand. Deep in a Madagascar jungle I was ambushed by a micro-assailant who liquefied my guts and could have drained me to a husk in a few days. Each time these predators attacked, I was able to get my hands on a tool that could beat them off. I'm fortunate.

In many cultures, young humans perish by the thousands after being attacked by gut-liquefying microbes. Carried in the feces of one human victim, these predators travel in water, dirty food, and on skin to the mouth of a new victim. They slide into the new prey, fuel up, produce a hundred generations in the blink of an eye, overwhelm the animal, and move on. A vast assortment of viruses, bacteria, and more complicated microbes make a living like this. It's a successful lifestyle, as long as humans cluster together in great enough numbers that micro-progeny can continue sliding out of one and into the next. And that has been the case for most humans ever since the invention of agriculture. The past ten thousand years have been very good for infectious predators.

Consider HIV, the virus that dismantles the human immune system. The virus enters a human and gets busy reproducing. Because viruses lack the major appliances found in most cells, they harness the prey's own cells and borrow those appliances. Thus equipped, they multiply. The cells that HIV prefers to hijack actually run the human's own immune system, and this means that the prey will eventually lose the ability to defend itself. But the predator HIV doesn't care. Even as it slowly digests this prey, it's reaching out for new animals to eat. Mixing with the prey's blood, milk, and semen, the virus bets that one day the prey will contact the vulnerable tissues of another human, and a few young viruses will carry the family DNA down the line. Yes, the original prey animal will die. But in a fresh-caught human, an HIV virus will carry the legacy forward.

All of this requires a certain density of human prey. If you, as an HIV virus, should capture a human that spends 99.9 percent of its time in the same small troop of humans, the odds of your dynasty outliving that group are iffy. Sure, you and your progeny can take down the whole pack. But if you eat the last one before it carries you into contact with another troop of prey, your run is over. Thank you for playing. You're the plague that never got started. Yours is not the strain of HIV that will win glory and headlines.

Sexual transmission, HIV's main means of hopping to new prey, is a common choice for microbial predators. Animals must reproduce—that's the mission nature issues to all living things. And in order to reproduce, most animals must mingle their sexual organs. That merger forms a bridge from one individual to the next. So adopting sexual transmission as your means of spreading is a no-brainer—which is ideal for organisms lacking a brain. But spreading via other body fluids can be efficient, too. The common cold, while not deadly enough to qualify as a predator, manipulates its host into releasing a river of mucus from the nose. The human can be relied on to wipe this mucus with its hands. When those hands contact the nose or mouth of another human, or even when they touch an object another human will touch, the virus lurking in the goo slithers into its next host. It's another savvy strategy.

Although the common cold is fairly innocuous, its fellow predators of the human lung rank on a par with the gut liquefiers, in terms of their kill rates. Across much of Africa, pneumonia is the top killer of humans under five years old. Many of the lung predators that cause pneumonia can hop from one victim to the next even without the humans even touching. They manipulate their current prey into coughing out an aerosol that's laden with the next generation. When fresh prey inhale the microscopic droplets, they, too, succumb to the predator's slow consumption of their lungs.

The most impressive micropredators, to my mind, are those that rely on a chain of unrelated species to complete their life cycles. The evolutionary steps involved in the disease of bilharzia, or schistosomiasis, for instance, are breathtaking. Embedded in human feces, eggs of the blood fluke *Schistosoma* enter a stream or pond. The eggs yield tiny organisms that drift until they encounter a snail. In the snail's liver they undergo their next metamorphosis, emerging as an army of free-swimming squirmies. When a human sets foot in this broth, the swimmers drill into its flesh, shedding their tails. Inside the human, they again hunt out the liver, where they convert human tissue to fully adult worms (causing anemia in the process). Male and female worms then mate. The female produces eggs that will exit the human gut along with feces. With any luck, they'll land near a body of water. . . .

Humans contend with a far greater number of microbial predators than animal ones. The list is long—anthrax, cholera, dengue, hepatitis, influenza, shigella, tetanus, tuberculosis, and yellow fever are a few. And the list is virtually endless, because new predators are evolving all the time. Recently, these include bird flu, SARS, Ebola, and HIV. Because a microbe can reproduce in minutes, it can evolve very quickly compared to a reptile or mammal that produces a new generation only once a year. Each round of microbe reproduction opens the door to genetic mutations that better equip the species for chasing down and consuming human prey. So, tiny as they are, our micropredators are far more lethal than the tigers and the bears.

Of course, the human can evolve in response to a microbial predator, just as we can to a furry carnivore. Our sluggish reproduction rate is a drawback when competing with a microbe. But eventually, we do find a way around the buggers. Etched in the DNA of some human races are dramatic tales of pursuit, capture, and escape. A whopping 10 percent of European humans harbor a mutation that blocks the entrance of HIV to their white blood cells. Why is this mutation so common in Europeans? Because earlier epidemics that swept through crowded Europe—perhaps smallpox or plague—killed off so many of the humans who didn't have that virus-blocking mutation. The fittest, those lucky mutants, survived. The nonmutants died in much greater numbers, amplifying the percentage of mutants in the population. These days a new virus, HIV, is picking off more nonmutants. Over time, the humans with the antiviral mutation will slowly, but surely, take over the world.

A PETTING ZOO OF PARASITES

The line between pure predators and parasites is a fuzzy one. One of the curious features of predation is that killing is not the predator's primary goal. Eating is. If lions could eat zebra meat without killing the zebra, that would suit them fine. They harbor no special hatred for zebras. They just need zebra flesh. And the same is true of the microbial predators. If the malaria bacterium can multiply and thrive while eating a live human, that suits malaria just fine. It's preferable, in fact, because a dead host may take its parasites down with it. So, I think of my predators as those who kill a large percentage of the humans they prey on, and my parasites as those who kill only a few, more or less by accident. The latter usually kill the young, the old, and the already sick. The classic parasite merely burdens its prey, eating it slowly enough to keep it alive.

Such creatures are extremely common in humans. Worms are currently supping on one in three humans worldwide. The tuberculosis bacterium is eating more than one out of every four humans in some parts of its range. In other parts of the world, malaria is dining on one in four children, many of whom will die.

The malaria protozoan—a one-celled beastie a tad more sophisticated than a bacterium—kills hundreds of humans for every one taken by crocodiles, for a total toll of between one and three million humans a year. But this is only a fraction of a percent of the humans it infects. Most humans preyed on by malaria maintain a long life span punctuated by bouts of illness as the parasite undergoes population explosions. Those humans who can't weather the periodic explosions, mainly children, perish. But malaria is one of those bugs that ought to prefer its prey to stay alive. This bug actually feeds on two animals, mosquitoes and humans. (Different species of malaria feed on all sorts of animals, but here we'll stick with our own persecutor.) Mosquitoes suck up the protozoan from one human, support it through a few of its dozen life stages, then inject ripe spores into the next human. Dead humans are not patronized by mosquitoes, and thus it is not in the interest of malaria to kill them.

The simple definition of a parasite is a plant or animal that steals from another, harming its host in the process. Parasites come in all shapes and sizes, ranging from the mistletoe plant to the lamprey fish, from fleas to the vampire bat. In my own life, I've served as prey for a variety of ticks, fleas, leeches, mosquitoes and other bloodsucking flies, and probably a worm or two. But again, my sanitized and tool-rich existence has protected me from the horde of parasites who chase down humans in other parts of the world. The animals who fill this niche are innumerable and often bizarre.

As a child I recall a family friend stopping in after a trip to Central

Asia. On her tanned shoulder a small volcano was growing. We all marveled at it, but she didn't see a doctor. A few weeks later her boyfriend lanced the thing, and fished out a chubby botfly larva. A variety of these flies prey on all manner of mammals, spending their larva-hoods eating flesh under the skin before erupting into winged maturity. Botflies are plenty icky. In Amazonia and the Orinoco River basin, humans think twice before swimming with the dreaded and diminutive candiru fish, who will wriggle into any undefended orifice, erect his spines, and stay a while dining on blood and meat. Across the Atlantic in Africa, the guinea worm makes its way down the human throat inside a tiny water flea who swims in unfiltered water. The female worm, after preying on the human for a year and mating with a handsome male guinea worm, roams the body until she's low in a leg. The blister she creates there drives the tormented human to soak in a cool pond, whereupon the worm protrudes her rear end to lay the eggs of a new generation. And of course humans host a swarm of skin parasites. Lice skulk around the hair roots. Chiggers burrow beneath the skin. Ticks and leeches cut a hole, lock in, and suck. Flies and mosquitoes dance in for a drink, dodge away, and dance back in.

The notion of all these mini-monsters preying on me puts my body in a new light: It's a habitat. Every inch of my corpus, inside and out, is valuable real estate to some microbe. My skin, even when it's not feeding mosquitoes and ticks, sustains a continuous film of bacteria. These are probably either mutualistic—we each derive a benefit from the relationship—or commensal—they benefit, and I'm unharmed. My innards are similarly settled: In my gut swarm between three hundred and one thousand species of bacteria, many of them helping me to dredge more nutrients from food. Upon investigation, I discover that I even have real-estate disputes raging within. Each species of gut microbe has a preferred acidity or a favored address, and each strives to elbow competing microbes aside.

The animals eating my innards can be much more aggressive than single-celled microbes in their territorial disputes, judging from studies on other animals. In rats, for instance, a tapeworm's meal is in jeopardy if a thorny-headed worm comes down the gut. The thorny-head likes the same habitat as the tape, and will drive it to a less advantageous position inside the rat prey. The animal annals also include a wasp species that lays two types of egg in living caterpillars, to produce both normal and "soldier" larvae. The soldiers hatch into warriors who patrol the inside of the caterpillar, doing battle with any competing parasites they encounter. (They die with the prey after their normal siblings mature and exit.) In my own gut, the most violent conflict I uncover involves battle between members of the same species. Once latched on to the gut or liver, some species of worm appear to cooperate with the prey (me) to defeat members of their own species who show up later. Like the lion pride that defends a fallen zebra from other lions, these worms murder one another to protect their meal. I should be flattered, I suppose. My body is a rich habitat, worth defending for all these animals and dozens upon dozens more.

How to fend them off, kill them, evict them? Against large carnivores I can level an arsenal of tools—stones, spears, bullets, traps, poisons. But traditional tools are impotent against the predators who would eat me from within. For interior defense, I have two paths of recourse. I have the standard mammalian immune system, and I have tools—piles and piles of medical tools.

My immune system specializes in identifying predators, then rallying inner resources to destroy the invaders. I can fire up the temperature of my entire body, which seems to both speed some immune reactions (the function of fevers is actually controversial), and to boil some invaders alive. At the drop of a hat, I can also forge customized cells programmed to seek and destroy micropredators. Often, these made-to-order defenses protect me for life, waiting in a weapons magazine, ready to smash any future invasion to bits. Thus one round of chicken pox is all I'll ever see. Every time that virus tries to enter again, my immune system is going to mobilize the correct platoon of antibodies to pulverize it. It's a pretty slick system, and it monopolizes a fair amount of my DNA. In fact, a recent estimate proposes that as much as 10 percent of my genome may be dedicated to customized warfare against internal predators.

And still it's insufficient. For one thing, the traditional predators—the colds, the flus, the diarrhea bugs—are always evolving. If they mutate sufficiently, a fresh generation can slip past the sentries in my defense system. But additionally, entirely new predators regularly pop into existence. Again, take HIV. No one knows how long the virus circulated among the chimpanzees of Cameroon, if indeed that's where it came from. But it does seem likely that by the 1930s the virus had evolved to defeat the human defenses. It was then that the disease probably embraced humans, probably starting with hunters who butchered and ate chimps. A few decades later, a similar strain named HIV-2 apparently underwent a similar mutation in sooty mangabey monkeys, and that, too, began to prey on humans. The genius of these predatory viruses is that they attack the immune system itself, greeting the cells sent to kill them with a knockout punch. And the HIVs are so novel to the human species, barely two generations old, that we've evolved no direct response to them. Like the naïve mammoths of North America when humans first appeared among them, we haven't had a chance to adapt. When HIV catches us, we just keel over and die—after helping it to hop to our closest friends, of course.

Ah, but we do have that ace in the hole, our medical tool kit. Again, HIV makes a fine example. In the space of twenty-five years humans dissected the virus's chemistry and behavior, and constructed a battery of weapons that can back the predator into a corner of the body and pen it there indefinitely. In fact, a tremendous amount of the human effort now bends itself to such challenges. With our carnivorous predators pacified, we're working feverishly to subdue the tiny ones, too. More than seven centuries have passed since a germ has been able to take down 10 or 15 percent of the global population. The Black Death spared Americans but mowed down one out of every six humans on the planet, and about one in two Europeans. The Americas were devoured a couple of centuries later by imported predators to which Europeans were somewhat immune. Smallpox and other microbes consumed 90 percent of some American races. Less than a century ago a newly mutated "Spanish flu" virus gorged on as many as fifty million humans, including my grandfather's beloved sister. However, by 1918 this represented only 2 percent of the world population. Humans were learning to defend ourselves.

But it would be misleading to suggest that we've just begun to fight the microbes. Humans may have grasped the germ theory of disease only in the past two centuries, but I'd bet we've been aiming our tools at those germs since we were tree-hopping primates. For some indication of the form ancient herbal medicine might have taken, we can ramble widely among the animal ranks. Chimpanzees medicate themselves with a variety of plants, from what we can tell. They'll even add clay to a mouthful of antimalaria leaves, which is proven to boost their potency. But much more "primitive" primates make medicine, too. A Costa Rican monkey chews up a suite of herbs to rub into its fur—the same herbs the local humans use to treat skin irritations.

Much more distantly related to us, the brown bear also chews a fragrant root to make a face wash. This cologne may function as an insect repellent. Miles and miles from us on the tree of life, even the spineless wooly bear caterpillar and the bird-brained European starling appear to employ plants medicinally. The wooly bear, when infested with parasitic fly eggs, prefers to focus its feeding on poison hemlock. The hemlock doesn't kill the fly larvae, who spend their first incarnation inside the caterpillar. But somehow eating hemlock helps the caterpillar survive when the parasites prepare for their next incarnation by chewing their way out through their host's skin. As for starlings, they line their nests with an assortment of plants proven to reduce the burden of lice on their chicks. A variety of other birds, including my local crows, treat their feathers with ants (alive or squashed in the bill), whose formic-acid secretions presumably repel parasites. A diversity of species this broad argues that the use of plant medicines is a behavior that has evolved many times over the eons. In other words, it's not that special.

How would such a thing evolve in an animal? How do the chimpanzee and the human both reach for the same plant—the mujonso, or bitter leaf tree—when they have diarrhea and gut parasites? The best guess is that all these animals, and perhaps many more we're unaware of, have a built-in ability to detect the chemicals they need. When their system is tipped off-kilter by a parasite, the body calls out for a certain compound. The nose then guides the animal to the tree, shrub, or herb that contains the right drug.

If other animals have been using plant medicines since heaven knows when, it's a reasonable bet that our hominid ancestors used them, too. Evidence will never be plentiful, because plants tend to rot quickly, leaving no fossil record. In the archaeological world, where a few old teeth constitute a press-stopping discovery, the likelihood of finding a million-year-old doctor's bag stuffed with *Echinacea* and St. John's wort is astronomically wee.

That said, in an ancient grave in Iraq scientists may have come close. Nine Neanderthals were buried by their kin in the Shanidar cave in eastern Iraq between sixty thousand and eighty thousand years ago. Neanderthal number four may have something to tell us about antimicropredator tools. Scientists routinely sample the soil around a fossil because the pollen there reveals which plants were growing nearby, and hence describes the ecosystem of the time. But the soil from Shanidar IV's grave contained more than a few stray pollen grains. It held clumps, as though entire blossoms had been buried with the body. Science's initial reaction was, "How sweet! They put flowers in their graves just as we do!" But there is something peculiar about the Shanidar flower species, each of which could be identified by its unique pollen shape. Almost all are still in use as medicinal plants.

As medicinal plants go, these were heavy hitters, too. Among the Shanidar plants was a species of *Ephedra*. Now, the types of *Ephedra* I'm familiar with produce flowers that resemble small, hairy fists. They're not beautiful. They don't scream, "We're so bereaved!" Much more interesting than its blossom is its namesake chemical, ephedrine. This does scream something: "Medicinal plant!" Ephedrine was an early asthma drug. Another flower was from the *Senecio* genus, and although the exact species isn't known, these plants contain chemicals similar to atropine, a workhorse of modern medicine that calms the heart and dries mucus. Again, the thistly blossom is unprepossessing. Also present was a yarrow, a lanky plant with a knobby flower, but whose sharp aroma hints at its chemistry. Long used in wound treatment, yarrow is an astringent that also fights swelling, stabilizes membranes, combats edema, and speeds clotting of blood. There was an *Althea* (mallow), whose charms include antimicrobial properties and an expectorant effect that would be useful against lung-invading micropredators. And there was a weedy, asterlike *Centaurea solstitialis*, an antiseptic and potent diuretic.

To my eye, all of these lack drama as cut flowers. So were these posies left for sentiment's sake, in which case the aesthetic tastes of the time were decidedly dowdy? Or were they an attempt to heal a patient in some afterlife? Or perhaps the tools of the dead man's trade, if he were his group's healer? Or, to throw a wrench in the works, was the pollen deposited in the grave by animals who were storing plants for a future meal? (It happens.) Unless and until we find more graves with more plants, we'll be left guessing.

Climbing up the ladder of time, we find the next fossil evidence of *Homo*s and medical tools, this time facedown in an Italian glacier. Ötzi, aka "the ice man," died about five thousand years ago, preserving a first-aid kit along with tools for hunting and building fire. Strung on a leather thong, he carried two chunks of bracket fungus, the sort of mushroom that grows out of a dead tree trunk. The species is something of a multipurpose drug, functioning as an anti-inflammatory, antibiotic, and a strong purgative. The latter could explain why scientists found parasitic worms in the process of being purged from their prey's colon. By Ötzi's time, acupuncture may have also entered the toolbox. His skin bore patterns of dark dots, simple tattoos made by rubbing charcoal or pigment into puncture wounds. The tattoos, around the spine, right knee, and left ankle, coincide with acupuncture spots in use today, and may have been meant to treat Ötzi's arthritis pain. Ötzi's medical sophistication is hardly surprising, just surprisingly complete in its exhibition.

By his time, all the world had dug deep into the medical tool chest. Just a few hundred years later, humans in China would compile a materia medica, written instructions for the use of nearly four hundred plants. And these days we know that humans exploit thousands upon thousands of plants against disease. This practice is not limited to dwellers in the rain forest and desert, either. Plants form the foundation of modern pharmacy, ranging from aspirin (from willow bark) to heart-saving digitalin (from foxglove), and cancer-fighting taxol (from a yew tree). In fact, two of the wildest frontiers in medical research today are "bioprospecting," in which scientists test random plants for useful chemicals, and "zoopharmacognosy," in which scientists save themselves a bundle of time by first taking note of which plants other animals are using.

The medical state of the art in Ötzi's time had also progressed to cutting and pasting of the body itself. Humans had moved far beyond the exploitation of plants, building more complicated medical techniques and tools to deal with wreckage left by predation attempts. By Ötzi's time, Egyptians were adept with sutures and would have used them to close wounds made by a carnivore or some other insult. (I suspect, however, that suturing is much older, as old as sewing itself, since any observant human would have noted the utility of the trick when applied to human skin.) By 2,500 years ago (again, at the very latest), a human who lost a body part to a predator had a shot at replacing it: At that date, Greek historian Herodotus related the case of a man who replaced his missing foot with a wooden substitute. Surgery, the art of cutting into a living human to repair damage, was also percolating at the time.

About two centuries ago the micropredators took their first body blow. Humans began toying with vaccination. This is the practice of borrowing weakened germs from one human or animal to infect another. The result is a set of those tailor-made warrior cells—immunity—gained without the inconvenience of a real infection. In 1853 the medical syringe arrived. Physicians used it to vaccinate everything in sight. (For those humans savaged beyond repair the syringe also sped the soothing sap of the Asian opium poppy—morphine.) And as the germ theory won converts, hand washing sentenced untold billions of germs to their deaths down a drain. The World Health Organization declared the smallpox virus extinct in 1980. Polio, reduced to a shred of a population, is on the brink. A small cloth filter suffices to strain guinea-worm larvae out of African drinking water. Screens and nets separate mosquitoes, with their malarial hitchhikers, from human prey. Sturdier shelters prevent polar bears and leopards from snatching humans in their beds. Machetes and mowing machines knock down the tall plants that hide crocodiles and alligators. And, of course, cheap bullets continue to mow down big carnivores when all else fails.

But we haven't won. About half the humans alive are currently under attack from at least one species of parasitic predator. Most of the predators are worms, which cause everything from generally lousy health to stunted growth, stunted intellect, elephantiasis, cancer, and blindness. And those crocodiles and leopards continue to harvest their thousands of pounds of flesh.

Furthermore, killing off our predators eases the limits on our own population. No animal does well when its population explodes. It tends to use resources too rapidly, foul its ecosystem with waste, and, ironically, pass around diseases more quickly.

What's more, the way human behavior has altered the planet's temperature is benefiting some of our worst tormentors. The World Health Organization estimates that the warming world already helps our smallest predators claim seventy-four thousand more humans each year than they did three decades ago. Many of the microbes that cause diarrhea and death when they feed in the human gut find heat most agreeable. Scientists tracking diarrhea rates in Peru have observed a rise in the rate of diarrhea that reflects El Niño's cyclical warming. As tropical temperatures spread out of their traditional bounds worldwide, the gut eaters will, too. Climate change allows these diarrhea bugs to kill an extra forty-seven thousand humans each year. Malaria takes down the rest. Warmth helps the malaria organism to mature much faster; it also hastens the mosquito's reproduction and spurs the insect to feed more. And, as with the gut bugs, malaria is hunting in new territories, as the gentle winters allow more and more mosquitoes to survive. The bacterium, which was brought to the Americas by European and African migrants, was driven from North America in the 1950s. But now it's resurgent, whining in Florida, humming in Texas, buzzing in Georgia. In Europe, it's returning to Italy, which had been safe since the 1970s. Even in its native tropics, malaria is spreading, climbing up mountains that were once too cool for mosquitoes. The predictions based on computer models of how the earth's surface will change show malaria swarming across Russia and the United States and digging deep into Australia in the coming century.

The degree to which *Homo sapiens* has altered its relationship with its predators is unprecedented in nature. It's not entirely clear how far the ramifications will ripple. Removing a single animal from an ecosystem can have a large, unforeseen impact. In this context, consider the wolf. Extirpated from the United States a century back, this predator formerly applied a brake to the ungulate population—elk, moose, deer. I can connect a few dots to argue that this predator once protected humans from Lyme disease, the tick-borne bacterium that attacks the human nervous system:

With wolves gone from the eastern forests, white-tailed deer were suddenly relieved of a serious predator. For about a hundred years, the human animal filled this role, holding the deer population under control. But as humans crowded into cities, our predation rate fell, and the deer rebounded. The tiny deer tick has never had such an easy time finding hosts. Its own population is booming along with the deer's. So when we defeated a large carnivore, we unwittingly opened a niche in the ecosystem for another enemy who's far more ruinous to human health than the gray wolf ever dreamed of being.

A similar dynamic is materializing in the micropredator arena. Vaccines, clean water, and hand washing have saved millions of human lives. But eliminating germs and worms has had two dark consequences. First, mounting evidence shows that the human animal is thoroughly adapted to carrying a load of parasites. Denied the balancing effect of worms or diarrhea bugs, many humans experience instead a major malfunction of their immune system: asthma, allergies, and Crohn's disease all appear to be calmed by a dose of micropredators.

Second, for many animals, predation applies a stern brake to population growth. This helps hold the creature to numbers the ecosystem can feed, water, and shelter. But humans have now disarmed so many of our predators that we're experiencing the transition to a different system of brakes. Instead of perishing from hyenas, the young humans of Ethiopia are dying from starvation. The young of Florida have thus far evaded malaria, but now crowding leads to aggressive conflicts that end with a more violent death. Throughout its range, the human animal is hitting limits—shortages of food, water, and healthy habitat.

Once again, the runaway brain of the Inventive Ape has transported its host animal to a context never previously constructed in nature. What befalls an animal who outwits its enemies, one after another? In many ways, it thrives—as we can plainly see. But this top-dog dynamic is only a few generations old. Already we're witnessing the effects of our unchecked reproduction. But some of the ripples that spread from a disrupted ecosystem travel quite slowly and are easy to overlook until they've become breaking waves.

So our unprecedented success as a prey animal is not total. Our ability to vanquish our enemies is a double-edged sword. Removing animals and microbes from our ecosystem changes the ecosystem—

on which our ultimate success depends. And even as we continue the campaign, we don't quite know where the second edge of the sword lies.

11

A BULL IN A CHINA SHOP:

ECOSYSTEM IMPACTS

The human, like every living thing, seeks to marshal the natural resources necessary to propel its offspring into the future. Just as the beaver kills trees and the termite emits carbon dioxide into the atmosphere, so does *Homo sapiens* alter the ecosystems wherein it resides. There the similarities end. For although beavers and termites are driven to control resources, humans have actually achieved this.

The animal's facility with tools is one factor in its dominance of natural resources. The human first strayed from a normal path of ecosystem alteration when it began to use tools to harvest prey animals. Many of these it hunted to extinction. Then, some ten thousand or fifteen thousand years ago, humans embraced digging tools with which they transformed many—in fact nearly all—farmable ecosystems. A few thousand years ago the animal's industrious behavior began to produce noxious gases and metals that have permanently altered the air and soil chemistry of the entire planet.

Two additional factors amplify the human's role in its ecosystem(s). One is the expansion of its range, which distributed the animal's ecosystem impacts across the globe in less than one hundred thousand years. And the primate's tremendous population, achieved through the defeat of most of its predators, multiplies its effects.

The entire planet, including its seas, atmosphere, and even space beyond, now bears the mark of this unusual creature. Its ecosystem impact, one could say, is total.

IN THE BEGINNING, WE ATE THE ANIMALS

Animal protein was the first food that ever passed my lips. It wasn't steak or eggs, but milk—milk of the female *Homo sapiens.* I've retained a taste for animal foods. I love the slick musk of melted cheese and the flowery brine of lobster. Like my grandmother, I could eat scrambled eggs three meals a day. Last night I had a charred and bloody hamburger that could only have been improved with a thatched roof of bacon. Recently I have eaten a slice of liver from a forcibly fattened duck.

I know this is not the most ethical way to meet my need for protein—especially the duck liver. Although I practice my predation on domestic animals now, one end result is the same as if I ate wild ones. The human's meaty diet imperils wild animals by appropriating their habitat for the production of grass and grain to feed the domestic chickens, pigs, and cows that I do eat.

When our prehuman ancestors relied mainly on plants for their nouriture, they blended in with the other apes. But on the day that the first pre–*H. sapiens* primate used a tool to slay another animal, we stepped onto a new path. The kill rate was going to go up. What day that was we'll never know, due to the nature of early implements. If that first prey animal was trapped, then the pit or net soon rotted into the background. If the hunter chose a spear, that, too, promptly disintegrated. Even if the original hunter wielded a durable stone, to an anthropologist's eye today that stone will look like a million others. Regardless, it must have happened: Proto-*Homo* took tool to animal. And its hunting success improved.

It probably didn't go up much, in those early eons. I once took a survival course in which students learned to use a "killing stick." For seven days I practiced whipping a stick at an unflinching piece of firewood, which stood just twenty feet away. At my peak, I only slew it one time in four. Even a professional baseball pitcher aiming for a target of about four square feet will succeed only about half the time, according to data at www.baseballanalysts.com. That's a big target (wolf size, roughly), at close range (sixty feet), but that pitcher's family is still going to go hungry half the time.

Very few hunter-gatherers remain who haven't adopted guns and other modern hunting tools, so I can find no studies of low-tech kill rates. But I imagine ours was a long and slow transition from a primate who ate plants and bugs to one who habitually stalked and killed bigger animals. Perhaps for a few million years meat was the exception to the vegetable rule. And among subsistence cultures today, meat remains a coveted treat, not a staple.

We can think of this middling predation rate as a baseline for *Homo sapiens.* When armed with minimal tools, we are a hunter more effective than a chimpanzee and less effective than a cheetah. When our tools and our population were both held at a minimum, we were presumably in balance with our prey. Like any predatory animal, early *H. sapiens* eked out a living by whacking his prey and dodging his predators. Nobody got fat. But enough humans got by to keep the species in the black.

AND THEN WE ATE LOTS OF THEM

But *Homo* expanded, as *Homo* will—both our toolbox and our range. The prey animals of Europe and Asia were prepared when *H. sapiens* migrated into their midst. Because *H. erectus* and *H. neanderthalensis* had already passed through, prey species had evolved an instinct that spurred them to flight when that smoke-stinking primate came tiptoeing around. But as *H. sapiens* crept onward, the creatures of Australia, New Zealand, Madagascar, and the Americas were caught by surprise. They had never met hominids before. Many of them were unable to evolve in time to save their species: A wave of extinctions followed the human around the world.

Many wild animals must learn their fear of a predator. This takes time. In the Galapagos Islands, a human can still sit down beside a sea lion on the beach, stroke its fur, and cut its throat. The Galapagos iguanas and penguins are equally innocent. (The fact that humans don't prey on those species slows their adaptation. Galapagos tortoises failed to evolve fear for a different reason: We ate them too quickly.) When the first *H. sapiens* arrived in Australia fifty thousand or sixty thousand years ago, it's safe to assume that the local citizenry were as naïve as modern Galapagos residents. Never mind that the natives then included a Komodo dragon–style lizard nearly twenty feet long, a six-hundred-pound kangaroo with spears for hind feet, and a marsupial the size of a rhino. Where humans were concerned, these giants were gentle. And short-lived. A few thousand years after *H. sapiens* arrived, one of us walked up to the last mega-ostrich and cut its throat. Humans had a new role in our ecosystems: eliminating other species. If we had accomplished this previously, it was a more occasional occurrence, and on a smaller scale. Now, as humans fanned across the globe, we etched our new role into the fossil record. In Australia alone, fifty-five large prey species blinked out of existence within a few thousand years of our arrival.

Next, the expanding humans invaded North America, striding in perhaps fifteen thousand or thirty thousand years ago. Within a few thousand years of our largest immigration, we ate most of that continent's large mammals, too. Giant bison, woolly mammoths, oversized antelopes, and colossal camels all proved defenseless against the new primate. One smaller species of bison did survive and adapt, only to face a second human offensive that would drive it to the very brink a few thousand years later. As we swaggered down through Central America we ran short of large mammals, who aren't common in tropical forests. But in South America our species wiped out a giant sloth in the Andes, along with more mastodons, saber-toothed cats, and an armadillo that weighed a ton or two. Madagascar was next to meet the modern man, who beached there about two thousand years ago. A two-hundred-pound lemur, the half-ton "elephant bird," and a small hippo began to sidle toward extinction. And then came New Zealand, where a fleet of flightless birds marked the human arrival by departing from the land of the living. (Islands like Madagascar and New Zealand tend to be mammal poor and bird rich, because birds can fly in to establish populations. Islands also breed enormous flightless birds, because flight is unnecessary on an island with few or no predators.)

The eraser of animals was a double threat because we not only used tools, but we were also studious enough to analyze unfamiliar prey when we encountered them. I retain this capacity now. If I had to live on the wildlife of my neighborhood, I'd use a different approach to catch each species. Squirrels, I've observed, will climb right into a box to get seeds. That's too easy. Raccoons are nocturnal, but they, too, will come to bait. The local skunk performs regular border patrols, so for him I'd wait by the porch at midnight. This scholarly approach to predation raised our lethality and sped our colonization of novel ecosystems. A cheetah, by contrast, is a one-trick pony. It chases, it swats, and it bites. An anteater who runs out of ants cannot switch to eating snakes. But armed with an understanding of its behavior, we humans can kill an animal ten times our size—sometimes without throwing a single spear. American bison were novel to the humans who arrived in North America fifteen thousand or twenty thousand years ago. They were 15 to 20 percent smaller than the extinct species, but there were millions upon millions of them. By exploiting the grazer's social structure and its pattern of stampeding when alarmed, humans could kill scores of them at a time. At the bottom of a Texas canyon, archaeologists have found a virtual hill of buffalo skeletons, left where they fell after humans spooked them into galloping off the cliff above. Humans chose the spot carefully, because the land sloped toward the canyon at precisely the right angle to make the drop-off invisible. They probably stationed themselves, along with rock-pile mock-ups of humans, on the prairie in a configuration that would funnel stampeding animals toward that sweet spot. And then they scared the herd into motion. In a single layer at the Texas site, it appears that eight hundred animals flew over the cliff before those behind could wrestle their instincts and their charging legs into a new course of behavior.

And that was without weapons. Overhunting was practically assured, given the human's ever-improving tool kit. If lions had guns, they would overhunt wildebeest and they would not pause to debate whether their harvest was sustainable over the long term. They would chow down and move on to the next-most-profitable animal. Actually, one predator has done this very thing, very recently.

The orcas of Alaska's Aleutian Islands have traditionally preyed on whales. But lately they began to take sea otters. Stumped by this—the otter is, relatively, a canapé—scientists analyzed the orca's diet and arrived at a hypothesis: The orcas, like us, have overfished one species after another. This started after humans ate whales nearly to the point of extinction after World War II. The Aleutian orcas turned to harbor seals, whose high fat content could fuel an orca for a day or two. The harbor-seal population plummeted. The orcas diversified to Steller sea lions, which are bigger but not as rich. The sea lion population plummeted. (Their own troubles were compounded by humans, who were competing for the fish they need.) And then the orcas shrugged and tried sea otters. They're not bad, but it takes six of them to fill you up. What the orcas will eat when the otters are gone the theorists don't propose. But like us, they're intelligent and adaptable animals with the behavioral flexibility to change hunting strategies as they change prey.

The human animal has traditionally displayed a similar disregard for population dynamics. This has been compounded by a surprising ignorance of animal reproduction. The otherwise-learned Greek philosopher Epicurus, for instance, proposed that to bring forth worms, one should combine manure, water, and sunlight. Fifteen hundred years later, a Welsh scholar pronounced that driftwood (fir logs, particularly) gave birth to the barnacle goose. As recently as the seventeenth century, European humans believed that mice could be bred from a jar containing wheat and sweaty underwear. (The sweat was thought to help the wheat to sprout.) And more recently, according to the late American Indian historian Vine Deloria, various North American cultures believed buffalo retreated underground in the winter, where they adjusted their numbers in response to human behavior before emerging the next spring.

Given our difficulty in grasping basic biology, it's understandable that the human animal is only now mastering the concept of a "last" mammoth or sea otter.

AND THEN OUR ANIMALS ATE EVEN MORE ANIMALS

There's a mouse in my house. There always seems to be. Even when they're not raiding my stored food, they come for shelter and perhaps water. My ecosystem impacts aren't all negative. Often, humans disrupt nature in a way that suits another species perfectly. Mice thrive in the human-dominated ecosystem. So do rats, pantry moths, body lice, houseflies, and a host of other species. As my human ancestors migrated around the globe, they transported with them a selection of these creatures. And this has added immensely to our affect on our ecosystems.

The human began transporting other animals a surprisingly long time ago. Some fifteen thousand or twenty thousand years ago in the western Pacific, humans transplanted a possum from one island to another. Whether they intended this, we don't know. Sailors probably provisioned their rafts with tree-trunk canteens of water, some seeds and fruit, and meat. Because dead meat spoils quickly in the tropics, sailors would have carried living meat in the form of possums.

So around they sailed, those ancient mariners, occasionally landing with a few animals still muttering in their cages. Again, whether humans released the animals for future use or the animals escaped, we can't say. But the animals spread with the humans. The first known transplant was the aforementioned possum, who was deposited on New Ireland, northeast of New Guinea. Then seven thousand years ago a different generation of humans outfitted that same island with a wallaby the size of a small dog. Next, pigs began boating about the western Pacific. Then chickens appeared in the trash heaps of Pacific islands. By Roman times, rabbits and sheep were in regular circulation around the Mediterranean. Each species introduced a new dynamic to its adopted home.

Besides these edible and companionable animals, humans accidentally gave passage to another class of creatures: the camp followers. My house mouse is one of these. Rats also proved to be formidable pioneers. Especially on islands, where birds have evolved to do much of their foraging and nesting on the ground, rats devastated the naïve natives. Suddenly, here was a predator whose omnivorous diet welcomed eggs and whose legendary wits were hardly taxed by picking them off the ground. On Pinzon Island in the Galapagos, every single baby tortoise to crawl forth from a wild egg was one eaten by a rat. (These days, humans intervene to save some eggs.) On the Caribbean island of Antigua, rats have extinguished the diminutive Antigua racer snake. And on the Pacific island of Midway, rats erased the Bulwer's petrel and nearly knocked off the Bonin petrel before humans repented, and extirpated the rats. Throughout the Pacific, the story of rats repeats.

Pigs also benefited from the human's expanding range. On Hawaii, pigs were set free first by Pacific humans and later by European ones. Together, they have transformed the native rain forest. They devour the low-growing plants, depriving birds of shelter and food. They furrow the earth into mucky wallows where mosquitoes breed and spread malaria among the endangered birds. In the course of rooting for food, pigs break open the earth for invading weeds and jump-start erosion, which sluices soil so far down the mountainsides that pig dirt actually smothers the coral reefs around the islands.

As with rats, the story of pigs gone wild repeats around the world. The explorer Hernando de Soto set some free in Florida in 1593. Those have spread throughout the South. More were unleashed on the fragile soils of California two hundred years later. And about one hundred years ago, humans decided to improve these free-running prey by crossing the feral hogs with European wild boar. The resulting race runs rampant now, gobbling native plants and crops alike and spreading erosion like a rash. And you can repeat that tale for South America, Australia, and New Zealand. On the Pacific islands where pigs took root, they usually shared a role with rats in the decimation of ground-nesting birds and of turtles, whose buried eggs they can smell and unearth. Just as mice thrive in my human shelter, so did pigs revel in the ecosystems that humans provided.

AND FINALLY, WE THOUGHT MAYBE WE'D BETTER STOP

The displacement of other species by *Homo sapiens* has been so sweeping that many of us now feel lonesome for other species. I'm a little fond of my mouse. I even tolerated a rat for a few days last winter, naming him Templeton as he regarded me from atop a bag of potatoes in the cellar. Prior to that I had a chipmunk who scampered indoors each day to harvest sunflower seeds from my desk. The human impact on other animals has rendered the world a little too empty for my taste.

Although I did eventually kill Templeton in a trap, I didn't kill him to eat him. I killed him in a competition over shelter, which he wanted to pee in. This symbolized another turning point in the human relationship to other animals: We kill other species proactively, before they can attack us or eat our food. This behavior spread along with a tool—the bullet—that made killing cheap and easy. Under fire in Europe, wolves and bears retreated to small patches of forest. When the bullet reached North America, humans made remarkably short work of predatory wolves, pumas, black and brown bears, coyotes, and even foxes that threatened domestic chickens, and crows that competed for crops. In Asia, tigers and leopards vanished, and in Australia the dingo dog and the bullet together doomed the Tasmanian wolf.

Better tools boosted our kill rate for prey animals, too. In North America, Indians upped their buffalo harvest first when they acquired horses from European migrants. They boosted it again when they got guns. When that world-altering tool of transportation, the locomotive, began to ply the Great Plains, European races of humans began killing buffalo by the thousand. They shot from the windows of their train cars, motivated by a hunger for the gush of brain chemistry that came from watching the crash of a huge animal. When tools make killing easy, humans kill like crazy: on land, on sea; animals both nutritious and noxious.

Trade also amplified the effect of the expanding humans. The passenger pigeon, whose migrating flocks could comprise two billion birds, is a textbook case. Indians had long eaten this North American bird. But the Indian population was no match for the pigeon population, and the sky-blotting flocks persisted. Even the early European migrants didn't much alter the dynamic. Each family of humans can eat only so many birds, after all. And although both Indians and Europeans were vigorous traders, recall that dead meat travels poorly. But when the locomotive poked its muzzle into the Midwest, the pigeon's prospects dimmed. For the first time, the market for pigeon meat extended beyond the humans who shared an ecosystem with the animal. Now when local humans had eaten their fill, they would net or shoot thousands more birds and stuff them into railroad cars. Hundreds of miles away, other humans would buy the birds for food or fertilizer. This sort of "predation by proxy" puts more pressure on an animal than its reproduction rate can cope with. We can see it today, with the hunting of Amazonian parrots for pets. Or the transport of gorilla meat to African cities, where humans eat wild animals out of hunger for approval from their social group.

Such headlong pursuit can't last. In recent decades humans have discovered that erasing an animal can come back to haunt us. The revelation is so new that it's still little known and less heeded. Nonetheless, experiments now argue that an ecosystem with many species of plants and animals is more stable and healthy than an ecosystem with less diversity.

We've also discovered a class of animals called "keystones." These are species without whom an entire ecosystem will crumble. The elephants of the African savanna are an elegant example. Their most transformative work is to clear brush. Shrubs and acacia trees are forever trying to gain a foothold in the grasslands, but elephants steadily devour them. Humans learned this in an accidental experiment, after we drove elephants off certain stretches of savanna. Acacias marched in, transformed the soil chemistry, and evicted the grasses. Along with the grasses went the zebras, gazelles, wildebeests, and other grazers.

So now we understand that the death of one species can reverberate throughout an ecosystem, possibly even rolling around to jostle . . . *us! Eek!* As yet, though, we don't know who all those keystone creatures are. They can appear very humble—the prairie dog, the beaver, and the sea otter are keystones. The otter, when not eaten by orcas, eats sea urchins. When those urchins aren't controlled by otters, they'll eat the kelp forests into oblivion. And when the otters are eaten by orcas, that's just what happens. The clear-cutting of kelp destroys the habitat required by young fish and a host of other creatures.

We have discovered such connections at a moment when one out of four surviving mammal species is at risk of extinction. A century ago, humans knowingly shot down the final flock of passenger pigeons, with an insouciance that would have looked well on an orca who was tossing down the last sea otter. But today, many humans feel differently.

Our sense of regret intrigues me. Any other animal on Earth would go on exploiting its fellow creatures until doomsday. The pig does not lose sleep over erosion in Hawaii. The black rat doesn't wring its paws about the bird eggs it has eaten. I'd bet my shelter that the orca doesn't find itself, in a quiet moment, reflecting on the moral dimensions of extinction.

I do. The human animal does. Or, to be absolutely accurate, the human animal has that capacity. Some of us (mainly we with secure food and shelter and a cultural proclivity) have an ability to contemplate the morality of our relationship to other species. And at that point, human behavior becomes unique and fascinating.

Only the human animal could agree that the panda bear deserves a series of reserves so it can survive in China. Only the human animal will climb aboard boats, go to sea, and risk its life to ram whaling vessels—not to claim the prey for ourselves, but to spare the whales from predation. Only humans will relinquish a powerful insect killer like DDT because it damages the eggs of eagles. Although a wild dolphin may occasionally intercede to protect a human from a shark attack—and it does appear they do—it's hard to imagine that the dolphin community has reached an agreement to adopt this policy at the expense of their own best interest.

It's a puzzling impulse, this desire to protect other animals from ourselves. It arose even before our awareness of how heavily the fate of our own ecosystems can depend upon a single species. So perhaps it's related to that puzzling human desire for pets. The fact that humans are quickest to save mammals suggests it could be self-love gone amok. It's harder, after all, to rally humans around a pinion moth than it is to mobilize in defense of a wide-eyed baby seal. For all we know, the moth is a keystone species. But it doesn't strum the human heartstrings the way a fuzzy face can.

Ultimately, humans who hope to spare other species will have to confront the fecundity of our own kind. My father liked to say, "There's no environmental problem today you can't solve with a 90 percent reduction in the human population." It's true. But who's volunteering to abstain from this selfish, dirty business of living?

Well, a few cultures, including that of China, have agreed that human reproduction must be limited by force (although this policy is for the protection of humans, not other creatures). And a smattering of individuals around the world have determined that the planet's need for fewer humans trumps their own desire to reproduce. These humans are represented by such organizations as the Voluntary Human Extinction Movement, whose website argues: "Phasing out the human race by voluntarily ceasing to breed will allow Earth's biosphere to return to good health."

And this, too, sets us apart. What other creature would deliberately thwart its reproductive instinct so that another animal might live? While it's a scintilla of a minority who entertain such thoughts, it is remarkable nonetheless: Only the human animal sacrifices so.

BACK IN THE BEGINNING WE ALSO REARRANGED THE EARTH

Today a human named Frank is scooping soil off my yard with a giant yellow machine and dropping it into a dump truck. My human family needs more room, and we're adding to our shelter. It took Frank a few minutes to uproot a forty-foot hedge of forsythia and a thirty-foot hedge of lilacs. The number of insects left homeless by this staggers my mind. And the woodpeckers who hunted those insects among the lilacs are cursing my name.

The human animal finds it useful to rearrange the face of the earth. Sure, other primates will dig a hole here, or break branches for a nest there. Sometimes a chimpanzee maddened by his own masculinity will go on a rampage that extends to busting up trees and shoving boulders into ravines. And primates have nothing on beavers. Seized by an instinct to squelch the sound of running water, a beaver will clear-cut a landscape of trees, whittle them to logs, and build dams. To a nest of orioles hanging in an oak tree by a stream, the beavers' yellow teeth must glint like the blade of a chain saw. To a colony of ants or a mouse with a nest full of blind infants, the rising water must look as dooming as the Yangtze River behind the Three Gorges Dam. All animals seek to bend their environment to the service of their own dear selves. The human animal just does it best.

As with animal hunting, the human ancestors got an early start in landscaping. Large-scale alterations probably began with fire, and fire remains as central to my own control of the landscape as it has been for any hominid. But where fire was once set free to sweep an area clean of plants, today it toils in a thousand different harnesses. A series of neatly timed blazes inside the engine of Frank's excavator serves to dig the hole where my shelter will expand. Elsewhere another flame consumes coal and powers the sawing of the timbers that will rise there. Half a continent away another fire cracks petroleum molecules into vinyl for my siding.

It's hard to say when we first tamed fire. It predates *Homo sapiens*, that much is certain. In various caves around the world, scientists have found charcoal and burned bones that suggest hominids controlled fire three-quarters of a million years ago—or as long ago as two million years. The data are few and (ahem) hotly debated. In any event, *Homo sapiens* matured as a committed arsonist. Sure, fire was great for both tenderizing and preserving meat and vegetables, and the snapping flames kept lions at bay. But *H. sapiens* also discovered how to rearrange an ecosystem by letting fire loose upon the land.

In all the accounts I've read of the ancient burning cultures—and they span the globe from the forests of Mexico to the African veldt—none matches the studious precision of the first Australians. According to interviews with modern Aborigines, a savanna with trees creeping in from the edges presents a sad and unkempt state of affairs. Fire is essential to chase out the trees and restore the tender plants that prey animals love. And so in northern Australia each clan has traditionally observed an annual cycle of burning to keep things tidy. As humans there migrated around their territory with the seasons, they left behind a careful patchwork of scorched earth that cleared the brush and renewed the soil. These were not big conflagrations (unless something went wrong), but small, surgical fires.

And they were transformative. Like the elephants who repel acacia trees on the African plains, many human groups have assumed the role of field keepers. In the forests of the North American East, prehistoric humans burned the understory to keep white-tailed deer both happy and visible. (The new growth fed the deer; its low stature revealed them.) Further west, a patch of prairie south of Lake Michigan persists where by all rights a forest should stand. New research speculates that human fire may explain both the creation and maintenance of this anomaly. And in the rain forests of Central and South America, preserved layers of charcoal and pollen grains show that early farmers there were slashing and burning great tracts of forest to grow crops at least seven thousand years ago. In short, many of the landscapes we treasure as untouched wilderness today were completely transfigured by humans.

Humans also deployed fire to clear trails through forest ecosystems. This opened the way not just for us, but also for a stream of other animals and plants that were formerly prevented from entering. The newcomers added a species here and subtracted one there.

With all this fire flying around, some of it on a schedule you could set your watch by, the mixture of plants and animals changed. As forests changed to fields, plants whose flammable seeds lay on the ground were slowly weeded out. On the other hand, plants whose pods or cones required fire to pop them open flourished. Birds and long-legged mammals were in luck; more languorous animals, like iguanas or sloths, saw their dreams go up in smoke. Where humans brandished fire, ecosystems metamorphosed. Interesting questions flare up when a landscape has been so altered by humans, and so many thousands of years ago: What lived here before? And what would you consider the "natural" landscape? Is it whatever was here before humans? Or is a human-made savanna just as natural as the beaver-dammed pond? Muddying the question further is the fact that farming humans, who migrated into the territories of hunter-gatherers in the past few centuries, often banned burning, which transformed those ecosystems yet again. A classic example: Since European agriculturists took northern Michigan from the fire-friendly Indians, the Kirtland's warbler has twittered toward extinction. It seems the bird will nest only in young jack pine trees, and jack pines only shed their seeds after a hot fire opens their cones. Clearly, when it comes to the landscape, we are a keystone species. The fortunes of millions have risen and fallen as the human animal adjusted the planet to suit our needs.

AND WE REARRANGED THE EARTH WITH GREAT DILIGENCE

I'm not content to scalp my backyard and leave it at that. After clearing the plants I don't want, I will import the ones I prefer. Out goes the lilac; in will come a tomato from South America, a basil plant from India, and a peony from China. Down the road at the farm where my summer vegetables grow, the transformation is more dramatic. An acre of corn, a field of pumpkins, and rows of spinach have replaced the native pines and oak trees. As the human animal bent its focus on farming, the surface of the earth began to change.

Humans may have been redistributing live plants for eons. Just as Pacific Islanders seeded new territory with meat animals, they also carried around wild food plants to set loose for future use. Plant foods have swirled around the planet so quickly that it's hard to recall that each underwent domestication in a particular ecosystem. Some surprises: Watermelon is not native to the American South, but to Africa. Ditto for okra. Cilantro is Middle Eastern, not Central American. The tomato is not Italian, but Andean. Sunflowers aren't French, but North American. Coffee is from North Africa, not South America. Hot peppers (and cool ones, too) hail from Central America, not India. Pineapples have their roots in South America, not the Pacific Islands. The orange (and lemon and lime) are hybrids of plants tamed in southeast Asia, not California.

And this plant-transport policy was no more "agriculture" than freeing wild possums was ranching. It was more like stashing cookies in your kitchen, your car, and at the office: You never know when you're going to need one, so you leave a few everywhere.

By ten thousand or fifteen thousand years ago, many of the world's cultures were routinely plucking and planting, plucking and planting, carrying plants on their journeys as casually as I might dig up a prized iris when I move from one shelter to another. Through agriculture, humans would eventually restructure 40 percent of the planet's earthen face. This is a stunning accomplishment, when you consider the size of the planet. It's not just that we've replaced oak trees with soy plants and cacti with cows. Our species swapping is only half the story. Along the way we've also tinkered with the very architecture of the earth. We've taken swamps and made them savannas. We've straightened rivers and filled in ponds. We've changed hillsides to stairways of terracing. The alterations even extend into the chemistry of soil.

Again, there's nothing unique about an animal digging up a little dirt. Moles do it. Foxes do it. In many parts of the world, fully half the soil owes its current structure to earthworms, who ingest loose particles but excrete robust crumbs. One ant species even practices horticulture, choosing, as the human does, which plants shall live and which shall die. The ant dwells in one Amazonian tree only, *Duroia hirsuta*, which has evolved hollow stems to shelter them. And it systematically poisons the trees that compete with *D. hirsuta* for sunlight. Like lawn fanatics with bottles of Roundup, the ants nip twigs, then inject formic acid, which kills the leaves.

But again, it's the scale of human efforts that makes our earthworks—rice paddies, brush-clearing fires, and banana plantations—visible from space. As humans worldwide adopted the new method of foraging, that is, bringing plants to the dooryard instead of hunting them down, we dedicated more space to our captive grasses, roots, fruits, and squashes. The reliable food supply boosted the reproduction rate, and our populations grew. That necessitated bigger gardens. This is when the serious earth moving began. In each ecosystem, it took a different form. In Mesopotamia, humans flattened and burned forests to create permanent fields. In the lowlands of South America, where wild squash surrendered to human desires some ten thousand or twelve thousand years ago, humans probably clear-cut new patches of forest in rotation, burning the slash to add a few years' worth of nutrition to the thin soils. In New Guinea, clear-cutting forests was just the start. There, by ten thousand years ago, large stretches of the soil itself had been carved into levees and ditches to maintain the moisture needed by yam and banana plants.

The transformation of Earth intensified further when animals fell under the human spell. Goats and sheep had accepted human dominance by eleven thousand years ago in the Middle East. What a tremendous relief it must have been to relegate the stalking of wild animals to an occasional "guys' weekend." These new meat animals stuck close to camp, and as a bonus they furnished milk as nutritious as their meat. And they were a great way to store extra carbohydrates: You could deposit your overripe fruits and moldy grains in a "goat bank," and eat them later, when you killed the goat and cashed in.

But these animals are *baaaad* for the land. Well, individually they're no worse than any wild mountain goat or Argali sheep who grazes in a dry ecosystem today. But when humans bring many of them together and move them infrequently, they chew the plant life into oblivion. Sheep tear at the turf, and goats defoliate young trees and bushes. Plants can take only so much abuse before their roots run out of gumption and they die. Then the soil once held by the roots is free to go. With the next rain or the next big wind, go it does. Pigs are even worse. When they forage they go straight to the root of the issue, using their nose hoes to flip up turf. Erosion is immediate. Cows, domesticated shortly after sheep, are gentler grazers, but their great weight crushes the crumbly texture of soil that allows it to breathe. Again, none of this would be a problem if the grazers were spread out like wild species. But we humans tend to keep all our animals within arm's reach. And that is what crushes the soil, starves the plants, and opens the earth to erosion.

Sometimes, looking at landscapes that we renovated in ancient times, it's hard to know which force, exactly, was the crucial one. The

Mediterranean hills of Greece and Italy, for instance, were once thick with pine and oak. Quite abruptly these ecosystems suffered both grazing livestock *and* busy humans, who cut the trees to use as fuel, to clear fields, and even to trade. When the dust settled, the landscape was completely new: It was still hilly, but shorn of trees, low on grass, and showing yellow soil. How did the three forces—clearing, grazing, timber trading—share responsibility for the resulting desert? Perhaps it doesn't matter, since all were traceable to the same ambitious ape.

Irrigation leaves a more obvious fingerprint on the land. Irrigation ditches made their first appearance in the arid lands of Mesopotamia eight thousand years ago. Once again, humans were seeking more efficient ways to forage: Instead of schlepping to the Tigris or the Euphrates to fetch water for your garden, why not dig a ditch and let the river come to you? Well, there are a few reasons why not. But these were so slow to reveal themselves that humans probably didn't notice a problem for two thousand years.

In Mesopotamia, irrigation at first brought a flowering of tools. In an agricultural system, a steady supply of food freed some humans from the daily hustle of feeding the family. Some could specialize in making new plows and tree choppers. Others could produce ceramic containers to keep surplus grain safe from mice. The human toolbox did run over. But as civilization took root, a couple of bogeymen lurked in the irrigated fields. Way upstream, both of Mesopotamia's rivers crossed modern Turkey, where other humans were busy turning forests into livestock pasture. The rivers loaded up with eroding soil. Then, when the silt-heavy waters diverged into the irrigation canals, they slowed and relinquished their Turkish burden. The canals slowly filled. Worse, when river water evaporates (instead of emptying into the sea), it leaves behind a trace of salt. In arid areas, this salt is never washed away. So each time Mesopotamian farmers sluiced the faintly saline water over their fields, a little more salt crystallized in the soil. The chemistry of the dirt would never be the same.

Year by year humans salted their soils. With such subtlety that no human would have noticed, salt-sensitive plants retreated. Decade by decade, century by century, wheat entered a sulk. More seeds failed to sprout—but not so many that any farmer would notice in her lifetime. The kernels on the spikes shrank as the wheat roots battled salt. But you wouldn't have remarked on this, looking at grains grown a decade apart. The alteration of the soil was slow but thorough. After a couple thousand years, wheat was practically extinct in Mesopotamian farmland. Barley, which tolerates higher salinity, filled the gap for a while, but the steady growth of the salt crystals eventually left much of the land unfit for grain plants.

Not all the human waterworks were self-destructive. One technique was so clever in its conception that modern humans are reviving it now in the South American Andes. From the air over Lake Titicaca, the land stretching back from the shore appears to be combed into narrow rows: Water, earth, water, earth. *Waruwaru*, the locals call this pattern, although until recently they had no idea why it existed. Patches of the corduroy propagate deep into the flat terrain, stretching the lake's water out over the land; it consumes more than three hundred square miles. The amount of labor this landscape represents makes my back ache. With stone and wooden tools, humans three thousand or four thousand years ago lifted chunk after chunk of soil from the canals and piled it onto the platforms. The platforms, we now know, were planted with potatoes and other food plants. But in this high, cool climate, the canals weren't for irrigation. They were perhaps the first human attempt at climate control. The water served as a radiator, soaking up solar heat in the daytime, then releasing it at night. So when the first frosts of autumn slid down the mountains and across the plains, the canals formed a warm moat around each platform. The growing season lengthened. What's more, the *waruwaru* drained the soggy earth, trapped eroding soil, and invited edible birds, fish, and snails into the garden. It's ingenious. If I had a handy lake, I'd go dig some *waruwaru* right now.

Downhill in Bolivia early humans seem to have used irrigation to farm fish. In this open landscape the rivers flood during the rainy season, inviting fish to swim over the savanna. And here humans built two hundred square miles of permanent ponds to capture floodwater and prolong the fishing season into the dry months. They also built spear-straight causeways of soil to connect their villages.

And back uphill in the Andes even above Lake Titicaca, the evidence of an early human makeover continues. The mountain slopes no longer bear a natural skin of foliage. Rather, they're sculpted, tip to toe, into terraces. Ancient humans found that flattening their gardens kept the soil in place, so they remodeled the mountains into staircases. This part of the Andes enjoys an image for being untrammeled and remote, but between the fish pens, the terraces, and the canals, it's about as wild as my driveway.

AND NOW WE ARE RECONSIDERING THESE RENOVATIONS

Where has the human's long history of land altering delivered us today? It's fair to say that by now the earth's very soil is an endangered species. After ten thousand years of accelerated erosion, healthy earth is getting scarce. The overall rate of erosion—from water, wind, and humans combined—is fifteen times higher than before humans evolved, according to one investigation. Between unintended erosion and the intentional alterations wrought by farms, roadways, shelters, public buildings, and cities, humans have now transformed between 50 percent and 83 percent of all the earth's surface. (Estimates vary.) Wild boar might tear up an acre of soil now and then, and beavers may drown a few snakes, but neither can present a serious challenge to the human record.

One fortunate feature of humans as our population grows is that our social disposition causes us to aggregate. So as humans bump against the limits of the land, we often maneuver ourselves into dense herds called cities. There, like social wasps or mound-building ants, we construct layered shelters that reach upward, not outward. While that pattern reduces the sprawl of human territories, it can affect the planet in other ways. Even if a human doesn't own food-growing territory, she still needs food. I now realize that my urban territory meets very few of my needs. While the squirrels in the backyard can wrest all their food and shelter from within the boundaries of their domain, I can't begin to support myself.

How much land would it take to prop me up—just to feed and shelter me, and furnish a new dress of fur or fiber once a year? This depends on the quality of the land, of course, and on which scientist is doing the calculations. But a few estimates clump in the range of a square kilometer of high-quality habitat per human. That's four-tenths of a square mile, or about 250 acres. In crummy habitat, like the Kalahari Desert, I might need eight times that, or more than three square miles. That's assuming I'm hunting and gathering for a living, not farming.

When humans began to cultivate plants, our territories shrank. Tame plants yield a lot of food in a small space. So even those humans who mix gardening and hunting, like the Amazonian Yanomami and some New Guinea tribes, still need less acreage than pure hunter-gatherers. How many acres does it take, then, to feed a human who's growing the bulk of her food but still hunting meat and cutting firewood? Less than seven acres of cleared land, according to one scholar's guess, plus a few acres of woodlot, and a hunting ground whose size would vary with the climate and the ecosystem. Certainly much less than 250 acres. As we tamed more plants and animals, we needed fewer and fewer acres to sustain ourselves.

With the industrialization of food, our "footprint," as ecologists call it, grew daintier still. As of fifteen years ago, the food eaten by the average Chinese human could be produced on two-tenths of an acre—precisely the size of my urban territory. I have often wondered whether my two-tenths could feed me, if I worked every inch. Apparently it could. But that wouldn't leave any room for a shelter, or trees for fuel. And besides, humans these days are nowhere near satisfied with food, fire, and a rude shelter. In fact, from the minute humans discovered the joy of metal tools, our footprint was destined to stray far out of the food patch. Metal is scattered through the ground in discrete patches. If your food patch boasts none, you'll have to either travel to dig it up or trade with someone who will dig it up for you. Most of your weight will continue to bear down in your food patch, but now one toe stretches miles away to the metal patch. And since discovering metal, humans have found a zillion other far-flung items we want, too.

My shelter, for instance, is far larger than necessary to keep me safe from wind and snow. The wood alone must have flattened a small stand of fir and maple trees. Stone to make the plaster probably came from a hundred miles up the coast. The nails and the plumbing may hail from many hundreds of miles west, where iron smelters were common a century ago. The coal that fired my shelter's original furnace would also have been Midwestern, and the first curtains at the windows were likely woven in the Southeast, near where the cotton grew. The human impact was spreading.

Today, my footprint dips a toe in every longitude around the globe. I can take a planetary tour without leaving my office. My feet are shod in clogs from Germany. The shirt hails from El Salvador, the sweater from Scotland. The phone is Malaysian, the microscope Russian, and the heater Italian. On the shelf stands a fish fossil from Wyoming, a green ceramic box from Iceland, and a red one from Thailand. And most of these items themselves contain elements and amalgams dug from the earth in a dozen more places. That's my office. The kitchen holds foods just as exotic: cheese from California; tofu from Massachusetts; curry and shrimp from Thailand; dried peppers from Mexico; fruit jam from Finland; wine from Australia; capers from Morocco. Each item carries with it a legacy of transportation. And its raw materials all had to be pulled out of one place, then processed in another. My big, fat hoof touches down in supertankers and cargo ships, aluminum mines and oil refineries, coconut groves and rubber plantations, Spanish olive trees and Russian glass grinders. It adds up.

Merely to feed me today requires four times the land needed by a Chinese human, whose diet is simpler and grown closer to home. (Well, that was the case in the early 1990s. China's desires are growing fast, and its environmental impact is, too.) The rest of my needs (assuming I'm an average American) commandeer another twenty acres, for a total of about twenty-four. This includes the fuels that produce heat, electricity, and transportation for me, the land covered by "my" cows, chickens, spinach, and cotton, and the forests for my paper and rayon, plus my water, and a dumping ground for my wastes.

Still, I could argue that my footprint is impressively small. A Stone Age farmer, whose worldly goods might have comprised a pair of sandals, a fire flint, a leather canteen, a fur cape, a few baskets, a net, a spear, a bow, a bear-tooth necklace, and a handful of stone blades and choppers, needed more land than I do because hunting wild animals is so inefficient. Modern humans can use land with meticulous economy. Even meat farms have become lean, mean, land-saving machines. In my culture, pigs live packed together, with each animal using very little space. Their own food grows in tight rows nearby, undiluted by trees and shrubs. Their manure, while copious, concentrates in ponds. If you freed the domestic hogs of the United States today, there would be devastation upon the land. In search of forage, sixty million hogs would root up forest and field, snarfling down bird eggs and trampling the soil into muck holes. Their own ecological hoofprint would splay across the nation.

On the other hand, most modern humans make a more modest impression than I do. The American foot is the biggest in the world. Our broad televisions, our towering shelters, our rotund cars, our wide-open sofas and bulging closets represent tentacles that tug on more land than the planet can currently afford. Most of the humans in China, Africa, and India rely on less than five acres' worth of planet, usually much less. True, they eat poorer foods, visit fewer dentists, and lead shorter lives than I will. But they're also getting along just fine without the two bathrooms, the ten throw pillows, and the twenty wineglasses that I've extracted from the planet's store of natural resources. Animals may be losing habitat faster in Africa than they are here in North America, but it's my giant footprint that's squeezing out those creatures.

Whether the human tendency to live in dense groups makes us more efficient isn't clear. A study of London in 2002 found that each human in that city had a footprint slightly bigger than the average of the entire country. But the Californian city of Santa Monica figures its rate to be twenty-one acres per human, smaller than the national average of twenty-four. This is a new science, and it will take time for analysts to perfect their methods. One gap in the formulas is that cities currently get no credit for services they render to outlying areas. Concentrated in cities are the world's hospitals and research centers, universities, financial centers, restaurants, libraries, and cultural trusts like museums—and most of the organizations in which humans search for ways to minimize our use of the planet. Just as the toes of a city reach deep into the surrounding farms and rivers, so does a rural area's toes extend into the cities.

Many aspects of urban life suggest it's more efficient. Urban territories and shelters are typically smaller, which limits the number of worldly goods each human can collect. For a sense of how compact a human city can be, imagine shaking the 8.2 million humans out of New York's five boroughs. Stacked as they are, each New Yorker now uses a territory averaging one-fiftieth of an acre. If you gave each of them a territory like mine, they'd cover ten times more land. From the five boroughs they crowd into now, they'd fill better than half of Connecticut. And without buses and subways, they would need a few million additional cars to achieve the national average of vehicles per household. Not that there's room for the New Yorkers in Connecticut. Connecticut, like most of the world, is already divided and conquered.

AND WE LET THERE BE LIGHT TWENTY-FOUR HOURS A DAY

Having adapted the planet's surface to our needs, the human proceeded to adapt the very atmosphere. The natural atmosphere was regrettably dark at night, and as soon as we could remedy that, we did. Rather abruptly two hundred years ago, humans tamed natural gas and set it afire to light the trails through our cities. In the 1930s electric streetlights replaced the gas lamps, and the night glowed brighter still. Today tamed photons mark our trails over much of the earth's inhabitable area, flaring at night into orange chrysanthemums over the suburbs and cities.

For the entire history of life on Earth, animals have been able to rely on nights being dark. Most have not adjusted. Studies of various birds, plus bats, mice, and other rodents, find that most species cut short their feeding and minimize travel if they're near lighted highways and streets. Some birds simply won't nest where lights intrude too brightly. After all, to potter about among the photons is to invite the attention of a predator. (For the same reason, many animals instinctively curtail their errands on full-moon nights.)

At worst, light can disorient animals. Insects, migrating birds, and baby sea turtles all fall under the spell of human lights. Insects circle the lights to exhaustion, as do night-migrating birds. Baby turtles mistake street light for the bright skyline above the dark ocean, and march into traffic. On the other wing, some bat species (but not all) benefit from our alteration of the air. It appears that they can catch bigger moths and make a night's meal more quickly, with the night-light on. In Florida, insect-eating geckos also seem adapted to the "street-light ecosystem." And the nocturnal skunks, raccoons, and possums whose territories overlap my own thrive in urban areas, heedless of glare.

Fortunately for the planet's other inhabitants, the human animal is able to undo some of our modifications, when we realize their effect on others. The humans in my culture once sucked water from the land without a second thought, but now we've all agreed to buy toilets that economize. Also in my culture we've agreed to halt any further alterations to ecosystems where endangered species dwell. And in a few tall buildings around the United States, a few humans voluntarily turn down the lights during migration season, to let birds pass through the territory unmesmerized. So, if the human's tools have magnified our impact on the planet, at least the same tools can now be turned to the purpose of replenishing the waters and the forests, and redarkening the skies.

AND ALSO WE DID POISON THE LAND, RIGHT FROM THE GET-GO

The yellow machine toiling outside my window draws its energy from captive fire. But having no use for fire's by-products, it sets them free: carbon dioxide, sulfur compounds, hydrocarbons in a hundred flavors. This happens a thousand times a day in my name, when I light the oven, when I shunt electricity through my clothes dryer, when I order my vinyl siding. This is the dark side of fire. It's our most powerful tool, but it's also our most poisonous.

When early humans cut a swath through the animals and furrowed the earth with canals and trenches, these modifications of the habitat were intentional. What came next, the poisoning of the planet, was not intended: Human life unavoidably produces by-products. Nor was it unusual. Plenty of creatures, and even plants, alter the chemistry of their ecosystems. The pine trees in my yard are busily acidifying the soil to their liking—and to hell with everyone else. Skunks when threatened will foul the air with sulfur compounds. And if it weren't for a certain strain of superpolluting bacteria, none of the above would be here at all: Three billion years ago blue-green algae accidentally poisoned the earth's atmosphere with oxygen, nearly putting themselves out of commission in the process. Life adapted, as life will, to the ruined air.

As with eradicating animals and reordering the landscape, hominids also got a surprisingly early start on polluting. They launched this career back when they first fooled with fire. Up until then, the dirtiest thing the hominids did was leave poop on the ground. But when they first burned wood, they commissioned a small cloud of dust and gases—mere irritants, plus toxins and carcinogens. This wasn't revolutionary stuff, since lightning started fires all the time. The point is that humans would one day rival lightning for the amount of smoke produced. And we'd set fire to all manner of nasty things besides wood, generating all manner of nasty smokes.

One of our earliest digressions with fire involved coal. This first occurred (that we know of) seventy thousand years ago at a site called Les Canalettes, in France. There, Cousin Neanderthal (presumably) was driven to innovate because an ice age was killing off the trees he had always used for fuel. Coal, like any other layer of rock, often finds itself folded and poking through the surface of the planet. Neanderthals discovered these rocks, and then discovered that they would burn. But unlike trees, coal contains abundant sulfur. And when coal burns, that sulfur rises into the air, reacts with moisture, and eventually returns to the earth as acid rain. Burning coal also releases toxic mercury and plentiful carbon dioxide, the chief cause of global warming. These by-products didn't amount to much until the Industrial Revolution, but I'm nonetheless impressed to find such an early start to our embrace of a fossil fuel.

Perhaps more insidious than the burning of coal was our melting of metals. Once a group of humans had settled in one place and begun to farm their food, then they could afford to work on new tools. Most human groups marched through their metal ages in a predictable pattern: copper, the easiest to melt, first, then its alloy bronze, then the most challenging (and rewarding), iron. In that light, it's not surprising that the earliest farmers were also the earliest metalworkers. Copper made its debut in central Turkey, a heartland of ancient farming, about ten thousand years ago. It took a few thousand years for the knowledge of this material to ripple, human to human, through the Middle East. Bronze follows; by the Roman era, Europe and Asia had graduated to iron. Also in a predictable pattern, toxic by-products poisoned the soil. Some of the old smelting works are still hazardous nine thousand years after their creation.

One of the most impressive is an ancient mine and smelting complex in modern Jordan. Wadi Faynan is a dry riverbed at the base of an arid mountain range. The soil here is still so thick with lead and copper that the plants and grazing goats carry massive burdens in their tissues. It's safe to assume that the humans who prey on these local plants and goats also bear a heavy load of metals. And this is thousands of years after humans first brought forth the poisonous metals from the ground.

The impact of the human ardor for metals wafted far beyond local forests and acid rains. When scientists sample cores from the glaciers of Greenland and the peat bogs of Switzerland they find that the by-products of our industrious nature went global. Because smelters discharged metal fumes into the air, the pollutants had a chance to mix in the atmosphere, traveling widely before they were washed out with rain and snow. Preserved in the buried layers of one Swiss bog, the concentration of lead begins a slow and steady climb about 3,200 years ago. Then in layers corresponding to the collapse of the Roman Empire and its smelting industry, the bog recorded cleaner air in the ensuing Dark Ages. And when the Industrial Revolution struck, the lead rained from the sky thicker than ever.

As I dig for studies of ancient pollution, I strike dioxins. These infamous chemicals made their name as defoliants in the Vietnam War, then became legends in papermaking and trash incineration. But a couple of UK biologists recently uncovered a colossal prehistoric source of the stuff. Dioxins can form whenever you burn something containing chlorine. Wood doesn't have much. Coal's content varies. But peat can hold buckets. Now, this isn't peat moss. This is an ancient and sodden layer of mosses, trees, dead bugs, sacrificed humans, and so forth, at the bottom of a bog. Dug out and dried out, peat makes a respectable fuel, as humans discovered many thousands of years ago. But peat near the ocean gets soaked in salt spray, and salt contains chlorine. A test burn of some fine Scottish turf yielded a shock: Two and three centuries ago, humans in the peaty Scottish Highlands and islands were positively blitzing the world with carcinogens. Per capita, they were forging dioxins fifteen to twenty-five times faster than do modern Brits.

As human industry percolated around the planet it demanded further uses of fire, which released further clouds of carbon dioxide. The human contribution of this climate-warming gas grew with each campfire, and then with each metal smelter and pottery kiln and charcoal kiln, each steam engine and electricity plant, each horseless carriage and giant coal-mining truck, and every backyard barbeque. Methane, another climate changer, also saw boom times starting about ten thousand years ago. Goats, sheep, and cows ferment their food in the gut, a process that produces methane. Each cow belch may seem like a piffling affair, but today the human animal's domesticated livestock are among the leading sources of methane. Happily, unlike blue-green algae, the human animal has been able to figure out that we're creating a problem for ourselves.

AND WE DID FIND THAT POISONED LAND BECOMETH POISONOUS

The net effect of the human animal on the planet has been indescribably huge. But if we're daunted by our own devouring of wildlife, shoveling of soils, and polluting of everything else, we are not defeated.

The water spurting forth from my tap is a prime example of how the human animal reacts when its environment turns hostile. Uphill at Sebago Lake, great changes have swirled through the water during the past hundred years. Perhaps a century ago, humans began building substantial summer shelters on its shores and depositing their feces in the sandy soil nearby. Next, the area's farmers stumbled onto the limit-busting miracle of fertilizers and pesticides, and these also found their way into the water. More humans built summer shelters, then the winter villages grew, and more feces oozed toward the lake underground. Vehicles leaked oil and gasoline onto the new roads. Cows splattered their methane-rich pies onto green fields. The water taps downhill no longer spurted clean water.

When blue-green algae poison themselves with oxygen, their only hope is to wait for a handy mutation that will help them adapt. When humans pollute ourselves into a pinch, we can invent a new tool. And so today the lake water passes through a treatment plant, where we strip it of hazards. Limits? Schmimits.

Thus we wriggle out from so many of the corners we back ourselves into. Running out of petroleum? Take this tiny plug of uranium and generate power from that. Is the ocean running out of fish? Build a fish farm. It's getting too expensive to purify aluminum and iron ore? Recycle the metals you've already purified.

Our tools aren't a panacea. With irksome predictability, a new tool spawns a new generation of challenges. The uranium is wildly efficient . . . until it's time to throw it away. The fish in the farms have to be fed . . . fish. The miracle of fertilizer forces twice the grain from an acre of land, which leads to . . . twice as many mouths to feed.

And that leads to the other problem with the human response to environmental limits. The only solution that stands a chance in the long haul is to limit our reproduction. Seven billion buffalo would eat themselves into a catastrophe. Seven billion orcas would, too. Seven billion dodo birds would have constituted just as sure a route to extinction as did the humans, rats, and pigs that killed them instead. The human animal has used tools to buy some time. But the laws of physics are inexorable. Sunlight and soil can build only so many plant cells. And plant cells can feed only so many animal cells. Ultimately, each of my animal cells is depending on plants to find the soil and sunlight it needs. As crumbling ecosystems shed brick after brick, each human cell is at the mercy of those keystone plants and animals who still—somehow, somewhere—hold the system up around us.

■ ■ ■

Homo sapiens has eaten dozens of its fellow animals into oblivion. We have dug up the soil and transformed the earth's surface so thoroughly that thousands more species have failed for loss of habitat. In our ravening for tools, tools, ever more tools, we have accidentally poisoned huge swaths of land, in addition to most of the freshwater and all—*all*—the air.

The modern human's impact on its environment is near total. My species has so mastered tools, and so inflated its population, that all the fun has gone out of Earth wrecking. It's too easy. We're capable of annihilating almost any species that bothers us. Our dominion over the land and its plants is challenged only by mountains and glaciers. We've even adjusted the planet's temperature—a rare feat indeed. And oddly, none of this Earth-altering behavior dates to our modern, messy era. The human animal has been messing up the planet since we first learned to strike flint.

Certainly the scramble to control territories and resources is not unique to the human. Foxes, given bulldozers for digging out bunnies, would make a stunning mess. Beavers, given truckloads of cement for their dams, would drown us all. But the human is the animal who has done it. The human caught on to tools, then rearranged its entire natural history around them. Now we dream up more of them every day.

The bad news about our species is that our progress from minor tinkerer, on a par with woodpeckers and beavers, to major transformer of Earth, has been blindingly fast. Woodpeckers have pecked wood for millions of years, declining to improvise or augment their wood-chipping technique. Beavers never toss a surprising new tree-felling or stream-stopping technique at their neighboring mice and orioles. It is the human animal who in the space of a few thousand years—an evolutionary snap of the fingers—has utterly transformed its behavior, and hence its impact on the planet. Now every species on Earth—not least the human animal—is challenged to adapt.

EPILOGUE

I hoped that defining my animal self would clarify my identity in the natural world. Having spent my entire life among wild and domestic animals, I have also spent a lifetime wondering how I fit in. I've been uncomfortable with the notion that I was an animal apart, a sort of extraterrestrial on my own planet.

Having cranked myself through the biologist's machinery, I see my animal self more clearly. I feel more personally the bond between me and a chimpanzee, and even between me and a fly. Our missions, after all, are identical. Every species is biologically programmed to escape predators and parasites, to gather food, to shelter from the weather, and to reproduce. Although each animal has evolved to meet these challenges in a different way, we are equals in the struggle.

In some ways my animal identity is a great comfort to me. I am still depressed by human warfare, greed, and oppression. But now it's clear to me that these nasty acts are as much a question of biological impulse as of personal scheming. Polar bears, baboons, and wild boar can be just as nasty and for just the same reasons. Somehow, the biological underpinning of human evil makes it easier for me to stomach. Furthermore, the human's great capacity for altruism and kindness shines even more brightly in the natural context. Our species is among the most generous and is clearly the most thoughtful.

It's that thoughtfulness that continues to alienate me from the rest of nature. My nerdy attention to the human's territorial behavior, diet, and fur color do help me to see where I fit in. But I'm afraid the human brain is always going to make my species stand out. It permits such a wide range of behaviors that there is simply no comparison in the natural world. We are, with regard to smarts, in a class of one.

I once had a dog that was half beagle, half Labrador. The soul of the animal was beagle, joyously bent on the acquisition of food. The body was Labrador. As a result, this was a beagle who could actually achieve goals most beagles only dream of. On those long legs he could reach a pie at the back of the kitchen counter or lift dishes out of the sink. It was a formidable combination.

The human brain is like the Labrador legs on that dog. In many ways, we're a normal primate. We yearn to control rich territories, build safe sleeping nests, and savor an occasional colobus-monkey steak. But with the brain that humans evolved, we can do all of this and so much more. Our primate urges and abilities are turbo-charged by that three-pound organ. We don't just capture territory, we totally transform it, removing forests or creating lakes as the need arises. We don't stop at building safe nests, we build bulldozers to accelerate nest building. And we're not satisfied to chase down prey animals with our bare bodies. We make tools to throw or shoot at them or pastures to contain them. As with a beagle on Labrador legs, nothing is out of our reach. And for most of our career as a species, we've been about as repentant as that dog was.

But lately, that spark of altruism and kindness has flared into an open flame of concern for the rest of the world's animals and even its plants. The human animal alone has the capacity to imagine the future, and we can envision the bleak landscape we'll inhabit if we continue to indulge our instinctive behaviors. Remarkably, the human also has the capacity to combat its own instincts. And that—the willingness to sacrifice for the other inhabitants of the planet—may be our most extraordinary quality. It is certainly the one that gives me the most hope for the future.

ACKNOWLEDGMENTS

I blew a lot of social capital on this book. Humans are obligate social animals—without social support, we'd die. We're also farsighted. So humans constantly make deposits and withdrawals in the big social bank, striving to build a reputation for paying back. But this wide-roaming book demanded a wide range of readers to hunt down errors and rein in the language. I've gone deep into the red.

My cousin Eleanor Holmes—forgive me, that's my *beautiful and brilliant* cousin Eleanor—is not just another mathematical genius with a pretty face. She did check every Fahrenheit-to-Celsius conversion, but she also corrected grammar and fearlessly proposed biological theories of her own, some of which turned out to be already proposed and proven. The brilliant and beautiful Jeff Steinbrink, scholar of the funnyman Mark Twain and quite an amusing writer himself, is humorless with regard to the comma—humorless and merciless in the persecution of the comma. Any commas remaining in this text are entirely my responsibility. Among the approximately eight hundred notes he made in the text was "The lion's share actually means 'all of it,' not 'most of it.' " You cannot pay for readers like mine with normal money. Only the guilt-weight of love can secure such services.

Margery Niblock, whom I've long worshipped for her proofreader's eye, is Jeff's equal regarding the hyphen. "My high school teacher told us to use them wherever they improve clarity," I protested. "Ach!" she spat. "It shows." Among her hundreds of notes was a reminder that Ping-Pong is capitalized. Uncle Tom Carper, a perfectionist poet, was true to form: A frequent formalist in poetry, he tested my every word for glibness, slang, and other distracting informalities. Aunt Janet Carper likewise tried to protect me from my own worst impulses and insisted that I was wonderful no matter what. Again, this quality of support can only be procured on the shadow market of human emotion.

The angelic Monica Wood, who knows well what it is to be mired in that dung pile we call a first draft, looked over the first chapter and said exactly what I needed to hear: It's so perfect that I should crawl out from under the desk and start fixing it. And that buying a new lipstick could help. The devilish Amy Sutherland, on the other hand, uncharacteristically kept her mouth shut until my suicidal phase passed. Our parallel book projects and e-dialogue provided that daily stimulation and irritation most work-at-homes only dream of:

HH: "Do you think Internet IQ tests inflate your score?"
AS: "What did you get?"
HH: "141."
AS: "Definitely."

My since-the-beginning reader and teacher extraordinaire, Kirsten Platt, wrote a critique of the early chapters so insightful and clear that I kept it on my desk for months. Nancy "Mom" Holmes, Stan Tupper, and Dan Lambert chimed in with fact checks that would occur only to the most congenitally curious humans. Sister Ellen recruited her biologist pal Grey Hayes, whose expertise on the destructive capacity of the human animal strengthened chapter 10.

My gratitude to the Ladies Toast and Boast Society remains boundless. My gratitude for librarians everywhere also defies quantification.

And my admiration for my mate, John Dorvee, continues to grow. Knowing nothing about writers, he married one. I, knowing something of his analytic capabilities, asked if he'd read. Proving that I really am the luckiest human alive, he, in addition to being patient, ridiculously understanding, and quite handsome, has also proven to be an instinctive and astute editor.

As I said, I am in serious debt.

Although none of the above can be had for hard currency, other services may be. I wouldn't want to navigate the weird waters of the book industry without the steady and intelligent guidance of my agent, Michelle Tessler. I'm grateful to Stephanie Higgs for selling this book to her colleagues at Random House, and delighted with Jill Schwartzman's willingness and ability to adopt it. Jill's input really did make this a better book!

SELECTED REFERENCES

The ease of literature searching hasn't quite eliminated the need for citations, but it's getting there. Here I include only the more interesting and obscure papers and texts.

1. QUICK AS A CRICKET: PHYSICAL DESCRIPTION

Ackerman, R. R., et al. 2004. "Detecting genetic drift versus selection in human evolution." *Proc Natl Acad Sci U S A* 101 (52): 17946–51.

Alexander, R. M. 2002. "Energetics and optimization of human walking and running: The 2000 Raymond Pearl Memorial Lecture." *Am J Hum Biol* 14: 641–48.

Auvert, B., et al. 2005. "Randomized, controlled intervention trial of male circumcision for reduction of HIV infection risk: The ANRS 1265 trial." *PLoS Biol* 2 (11): e298.

Bramble, D. M., et al. 2004. "Endurance running and the evolution of *Homo.*" *Nature* 432: 345–52.

Castellsague, X. 2005. "Chlamydia trachomatis infection in female partners of circumcised and uncircumcised adult men." *Am J Epidemiology* 162 (9): 907–16.

Cavalli-Sforza, L. L. 1986. *African Pygmies.* Orlando, Fla.: Academic Press.

De Waal, F. B. M., et al., eds. 2003. *Animal Social Complexity.* Cambridge, Mass.: Harvard University Press.

Marks, J. 2005. *What It Means to Be 98 Percent Chimpanzee: Apes, People, and Their Genes.* Berkeley: University of California Press.

Pond, C. M. 1998. *The Fats of Life.* Cambridge, UK: Cambridge University Press.

Relethford, J. 2005. *The Human Species: An Introduction to Biological Anthropology.* New York: McGraw-Hill.

Sarringhaus, L. A., et al. 2004. "Bilateral asymmetry in the limb bones of the chimpanzee (*Pan troglodytes*)." *Am J Phys Anthropol* 128 (4): 840–45.

Soulsby, E. J. L., et al., eds. 2005. "Sporting injuries in horses and man: A comparative approach." Havemeyer Foundation Monograph Series no. 15. R&W Communications, Suffolk, UK.
Spielman, R., et al. 2007. "Common genetic variants account for differences in gene expression among ethnic groups." *Nat Genet* 39: 226–31.
Stanford, C. B. 2006. "Arboreal bipedalism in wild chimpanzees: Implications for the evolution of hominid posture and locomotion." *Am J Phys Anthropol* 129: 225–31.
Townsend, C. R., et al., eds. 1981. *Physiological Ecology: An Evolutionary Approach to Resource Use.* Sunderland, Mass.: Sinauer Associates.

2. CRAFTY AS A COYOTE: THE BRAIN

Calvin, W. H. 2004. *A Brief History of the Mind.* New York: Oxford University Press.
Deary, I. J. 2006. "Genetics of intelligence." *Eur J Hum Genet* 14: 690–700.
Falk, D. 2004. *Braindance: New Discoveries about Human Origins and Brain Evolution.* Rev. ed. Gainesville: University Press of Florida.
Falk, D., et al. 2005. "The brain of LB1, *Homo floresiensis.*" *Science* 308 (5719): 242–45.
Kelly, B. D. 2004. "Neurological soft signs and dermatoglyphic anomalies in twins with schizophrenia." *Eur Psychiatry* 19 (3): 159–63.
Kopiez, R., et al. 2006. "The advantage of a decreasing right-hand superiority: The influence of laterality on a selected musical skill (sight reading achievement)." *Neuropsychologia* 44 (7): 1079–87.
Manning, J. T. 2002. *Digit Ratio: A Pointer to Fertility, Behavior, and Health.* Brunswick, N.J.: Rutgers University Press.
McManus, C. 2002. *Right Hand, Left Hand: The Origins of Asymmetry in Brains, Bodies, Atoms, and Cultures.* Cambridge, Mass.: Harvard University Press.
Relethford, J. 2005. *The Human Species: An Introduction to Biological Anthropology.* New York: McGraw-Hill.

Wilson, F. R. 1998. *The Hand: How Its Use Shapes the Brain, Language, and Human Culture.* New York: Pantheon.

3. BLIND AS A BAT: PERCEPTION

Alzenberg, J., et al. 2001. "Calcitic microlenses as part of the photoreceptor system in brittle stars." *Nature* 412: 819–22.
Barrett, H. C., et al. 2005. "Accurate judgments of intention from motion cues alone: A cross-cultural study." *Evol and Hum Behav* 26: 313–31.
Chandrashekar, J., et al. 2006. "The receptors and cells for mammalian taste." *Nature* 444: 288–94.
Davidson, R. J., et al. 2004. "Asymmetries in face and brain related to emotion." *Trends Cogn Sci* 8 (9): 389–91.
De Waal, F. B. M., et al., eds. 2003. *Animal Social Complexity.* Cambridge, Mass.: Harvard University Press.
Erzurumluoglu, D. S. 2003. "Sex and handedness differences in eye-hand visual reaction times in handball payers." *Int J Neurosci* 113 (7): 923–29.
Fay, R. R., et al., eds. 1994. *Comparative Hearing: Mammals.* New York: Springer-Verlag.
Owings, D. H., et al., eds. 1998. *Animal Vocal Communication: A New Approach.* Cambridge, UK: Cambridge University Press.

4. FREE AS A BIRD: RANGE

Barnett, S. A., et al. 1989. "Wild mice in the cold: Some findings on adaptation." *Biol Rev Camb Philos Soc* 64 (4): 317–40.
Beall, C. M. 2001. "Adaptations to altitude: A current assessment." *Annu Rev Anthropol* 30: 423–56.
Bridges, E. L. 1949. *Uttermost Part of the Earth.* New York: E. P. Dutton and Co.
Ceballos G., et al. 2005. "Global mammal conservation: What must we manage?" *Science* 309: 603–07.

Farkas, L. G., et al. 2005. "International anthropometric study of facial morphology in various ethnic groups/races." *J Craniofac Surg* 16 (4): 615–45.
Jurgens, K. D., et al. 1988. "Oxygen binding properties, capillary densities and heart weights in high altitude camelids." *J Comp Physiol* [*B*] 158 (4): 469–77.
McElroy, A., et al. 1989. *Medical Anthropology in Ecological Perspective.* Boulder, Colo.: Westview Press.
Post, P. W., et al. 1975. "Cold injury and the evolution of 'white' skin." *Hum Biol* 47 (1): 65–80.
Schaefer, O., et al. 1974. "Regional sweating in Eskimos compared to Caucasians." *Can J Physiol Pharmacol* 52 (5): 960–65.
Shea, B. T. 1977. "Eskimo craniofacial morphology, cold stress, and the maxillary sinus." *Am J Phys Anthropol* 47 (2): 289–300.
Sol, D., et al. 2002. "Behavioral flexibility and invasion success in birds." *Anim Behav* 63: 495–502.
Steegman, A. T. 2005. "Cold response, body form, and craniofacial shape in two racial groups of Hawaii." *Am J Phys Anthropol* 37 (2): 193–221.
Stinson, S., et al. 2000. *Human biology.* Wilmington, Del.: Wiley-Liss.
Thompson, E. E., et al. 2004. "CYP3A variation and the evolution of salt-sensitivity variants." *Am J Hum Genet* 75: 1059–69.

5. A DOG IN THE MANGER: TERRITORIALITY

Ardrey, R. 1970. *The Territorial Imperative: A Personal Inquiry into Animal Origins of Property and Nations.* New York: Atheneum.
Hamilton, M. J., et al. 2007. "Nonlinear scaling of space use in human hunter-gatherers." *Proc Natl Acad Sci U S A* 104 (11): 4765–69.
Hicks, D., et al. 2001. "Preventing residential burglaries and home invasions." International Centre for the Prevention of Crime. http://www.crime-prevention-intl.org.
Itô, Y. 1978. *Comparative Ecology.* Cambridge, UK: Cambridge University Press.

McLachlan, M., et al. 1990. "A conceptual model of organic chemical volatilization at waterfalls." *Environ Sci Technol* 24 (2): 252–57.

Taylor, R. B. 1988. *Human Territorial Functioning: An Empirical Evolutionary Perspective on Individual and Small Group Territorial Cognitions, Behaviors, and Consequences.* Cambridge, UK: Cambridge University Press.

Thorpe, W. H. 1974. *Animal Nature and Human Nature.* Garden City, N.Y.: Anchor Press.

6. HUNGRY AS A WOLF: DIET

Aguirre, A. A., et al. 1999. "Descriptive epidemiology of roe deer mortality in Sweden." *J Wildl Dis* 35 (4): 753–62.

Berger, J., et al. 2004. Remarks on moose mortality in Wyoming. From Wyoming Fish and Game Commission meeting, September 9, in Casper, Wyo.

Cordain, L., et al. 2002. "The paradoxical nature of hunter-gatherer diets: Meat based yet non-atherogenic." *Euro J Clin Nutr* 56 (Supp1. no. 1): S42–S52.

Felton, J. S. 1995. "Food mutagens: Mutagen activity, DNA mechanisms, and cancer risk." *Sci and Technol Rev*, http://www.llnl.gov/str/pdfs/09_95.pdf.

Gebhardt, S. E., et al. 2002. "Nutritive value of foods." *USDA ARS Home and Garden Bulletin* 72.

Hayes, M., et al. 2005. "Low physical activity levels of modern *Homo sapiens* among free-ranging mammals." *Int J Obes* 29: 151–56.

Hill, K., et al. 2001. "Mortality rates among wild chimpanzees." *J Hum Evol* 40: 437–50.

Kerlinger, P., et al. 1988. "Causes of mortality, fat condition, and weights of wintering snowy owls." *J Field Ornithol* 59 (1): 7–12.

Milton, K. 2000. "Hunter-gatherer diets—a different perspective." *Am J Clin Nutr* 71: 665–67.

Olshansky, S. J. 2005. "A potential decline in life expectancy in the United States in the 21st century." *N Engl J Med* 352: 1138–45.

Paquet, P. C., et al. 2001. "Mexican wolf recovery: Three-year program review and assessment." Conservation Breeding Specialist Group, for U.S. Fish and Wildlife Service.

Pond, C. M. 1998. *The Fats of Life*. Cambridge, UK: Cambridge University Press.

Speth, J. D. 1989. "Seasonality, resource stress, and food sharing in so-called 'egalitarian' foraging societies." *J Anthopol Archaeol* 9: 148–88.

Sponheimer, M., et al. 1999. "Isotopic evidence for the diet of an early hominid, *Australopithecus africanus*." *Science* 283: 368–70.

Ungar, P. 1998. "Dental allometry, morphology, and wear as evidence of diet in fossil primates." *Evol Anthropol* 6 (6): 205–17.

Wrangham, R. W., et al. 1999. "The raw and the stolen." *Curr Anthropol* 40 (5): 567–94.

Yamauchi, T., et al. 2000. "Nutritional status, activity pattern, and dietary intake among the Baka hunter-gatherers in the village camps in Cameroon." *Afr Study Monogr* 21 (2): 67–82.

7. LOOSE AS A GOOSE: REPRODUCTION

Anderson, K. G., et al. 2001. "Men's financial expenditures on genetic children and stepchildren from current and former relationships." University of Michigan: Population Studies Center Research Report no. 01–484.

Arnqvist, G., et al. 2005. *Sexual Conflict*. Princeton, N.J.: Princeton University Press.

Birkhead, T. 2000. *Promiscuity: An Evolutionary History of Sperm Competition*. Cambridge, Mass.: Harvard University Press.

Blurton-Jones, N., et al. 2002. "Antiquity of postreproductive life: Are there modern impacts of hunter-gatherer postreproductive life spans?" *Am J Hum Biol* 14: 184–205.

Buller, D. J. 2005. *Adapting Minds: Evolutionary Psychology and the Persistent Quest for Human Nature*. Cambridge, Mass.: MIT Press.

Burriss, R. P., et al. 2006. "Effects of partner conception risk phase on male perception of dominance in faces." *Evol and Hum Behav* 27 (4): 297–305.

De Waal, F. B. M., et al. 2003. *Animal Social Complexity.* Cambridge, Mass.: Harvard University Press.

Fisher, H. E. 1987. "The four-year itch." *Nat Hist* 96 (10): 22–29.

Foerster, K., et al. 2003. "Females increase offspring heterozygosity and fitness through extra-pair matings." *Nature* 425: 714–17.

Forbes, S. 2005. *A Natural History of Families.* Princeton, N.J.: Princeton University Press.

Forsyth, A. 1986. *A Natural History of Sex.* New York: Charles Scribner's Sons.

Garamszegi, L. Z., et al. 2005. "Sperm competition and sexually size dimorphic brains in birds." *Proc Biol Sci* 272 (1559): 159–66.

Grub, W. B. 1914. *Unknown People in an Unknown Land: An Account of the Life and Customs of the Lengua Indians of the Paraguayan Chaco, with Adventures and Experiences During Twenty Years' Pioneering and Exploration Amongst Them.* London: Seeley, Service & Co.

Hrdy, S. B., et al. 1984. *Infanticide: Comparative and Evolutionary Perspectives.* New York: Aldine.

Kappeler, P. M., ed. 2000. *Primate Males: Causes and Consequences of Variation in Group Composition.* Cambridge, UK: Cambridge University Press.

Lahdenperä, M., et al. 2004. "Menopause: Why does fertility end before life?" *Climacteric* 7: 327–32.

Little, A. C., et al. 2002. "Partnership status and the temporal context of relationships influence human female preferences for sexual dimorphism in male face shape." *Proc R Soc Lon B Biol Sci* 269: 1095–100.

McElroy, A., et al. 1989. *Medical Anthropology in Ecological Perspective.* Boulder, Colo: Westview Press.

Milius, S. 1998. "When birds divorce: Who splits, who benefits, and who gets the nest." *Sci News Online*, http://www.sciencenews.org/pages/sn_arc98/3_7_98/bob1.htm.

Moore, M. 1998. "Nonverbal courtship patterns in women: Rejection signaling – an empirical investigation." *Semiotica* 118: 201–14.

Reichard, U. H., et al. 2003. *Monogamy: Mating Strategies and Partnerships in Birds, Humans, and Other Mammals.* Cambridge, UK: Cambridge University Press.

Roberts, S. C., et al. 2004. "Female facial attractiveness increases during the fertile phase of the menstrual cycle." *Proc R Soc Lon B Biol Sci* 271: S270–S272.

Small, M. F. 1993. *Female choices.* Ithaca, N.Y.: Cornell University Press.

Stiffman, M. N., et al. 2005. "Household composition and risk of fatal child maltreatment." *Pediatrics* 109: 615–21.

Stinson, S., et al. 2000. *Human Biology.* Wilmington, Del.: Wiley-Liss.

Temrin, H., et al. 2000. "Step-parents and infanticide: new data contradict evolutionary predictions." *Proc R Soc Lon B Biol Sci* 267: 943–45.

Voland, E., et al. 2005. *Grandmotherhood: The Evolutionary Significance of the Second Half of Female Life.* New Brunswick, N.J.: Rutgers University Press.

Wasser, S. K. 1983. *Social Behavior of Female Vertebrates.* New York: Academic Press.

Zeh, J. A., et al. 2006. "Outbred embryos rescue inbred half-siblings in mixed-paternity broods of live-bearing females." *Nature* 439: 201–03.

Zerjal, T., et al. 2003. "The genetic legacy of the Mongols." *Am J Hum Genet* 72: 717–21.

8. BUSY AS A BEAVER: BEHAVIOR

Alford, J. R., et al. 2005. "Are political orientations genetically transmitted?" *Am Polit Sci Rev* 99 (2): 115.

Bekoff, M., et al., eds. 1998. *Animal Play: Evolutionary, Comparative, and Ecological Perspectives.* Cambridge, UK: Cambridge University Press.

Bowles, S. 2006. "Group competition, reproductive leveling, and the evolution of human altruism." *Science* 314: 1569–72.

Buller, D. J. 2005. *Adapting Minds: Evolutionary Psychology and the Persistent Quest for Human Nature.* Cambridge, Mass.: MIT Press.

Burnham, T. C., et al. 2005. "The biological and evolutionary logic of human cooperation." *Analyse & Kritik* 27: 113–35.

Delgado, M. R. 2005. "Perceptions of moral character modulate the neural systems of reward during the trust game." *Nat Neurosci* 8 (11): 1611–17.

De Waal, F. B. M., et al. 2003. *Animal Social Complexity.* Cambridge, Mass.: Harvard University Press.

Dugatkin, L. 1999. *Cheating Monkeys and Citizen Bees: The Nature of Cooperation in Animals and Humans.* New York: Free Press.

Fallon, J. H., et al. 2004. "Hostility differentiates the brain metabolic effects of nicotine." *Cogn Brain Res* 18: 142–48.

Gintis, H., et al., eds. 2005. *Moral Sentiments and Material Interest: The Foundations of Cooperation in Economic Life.* Cambridge, Mass.: MIT Press.

Griffiths, R. R., et al. 2006. "Psilocybin can occasion mystical-type experiences having substantial and sustained personal meaning and spiritual significance." *Psychopharmacology* 187 (3): 268–83.

Gurvan, M., et al., 2006. "Determinants of time allocation across the life span." *Hum Nat* 17 (1): 1–49.

Guthrie, R. D. 2006. *The Nature of Paleolithic Art.* Chicago: University of Chicago Press.

Keeley, L. H. 1996. *War Before Civilization.* New York: Oxford University Press.

Lewis-Williams, D. 2002. *The Mind in the Cave: Consciousness and the Origins of Art.* London: Thames & Hudson.

Nettle, D. 2005. *Happiness: The Science Behind Your Smile.* Oxford, UK: Oxford University Press.

Niebauer, C. L., et al. 2002. "Hemispheric interaction and consciousness: Degree of handedness predicts the intensity of a sensory illusion." *Laterality* 7 (1): 85–96.

Niebauer, C. L., et al. 2004. "Hemispheric interaction and beliefs on our origin: Degree of handedness predicts beliefs in creationism versus evolution." *Laterality* 9 (4): 433–47.

Sapolsky, R. M. 2004. "Social status and health in humans and other animals." *Annu Rev Anthropol* 33: 393–418.

Sussman, R. W., et al. 2005. "Importance of cooperation and affiliation in the evolution of primate sociality." *Am J Phys Anthropol* 128: 84–97.

Westen, D., et al. 2004. "Neural bases of motivated reasoning: An fMRI study of emotional constraints on partisan political judgment in the 2004 U.S. presidential election." *J Cogn Neurosci* 18 (11): 1947–58.

Wrangham, R. W., et al. 2006. "Comparative rates of violence in chimpanzees and humans." *Primates* 47: 14–26.

9. CHATTY AS A MAGPIE: COMMUNICATION

Arnold, K., et al. 2006. "Semantic combinations in primate calls." *Nature* 441: 303.

Bower, B. 2000. "Building blocks of talk: When babies babble, they may say a lot about speech." *Sci News* 157 (22): 344.

Bradbury, J. W., et al. 1998. *Principles of Animal Communication.* Sunderland, Mass.: Sinauer Associates.

Calvin, W. H., et al., eds. 2000. *Lingua ex Machina: Reconciling Darwin and Chomsky with the Human Brain.* Cambridge, Mass.: MIT Press.

Colantuoni, C., et al. 2002. "Evidence that intermittent, excessive sugar intake causes endogenous opioid dependence." *Obes Res* 10 (6): 478–88.

De Waal, F. B. M., et al. 2003. *Animal Social Complexity.* Cambridge, Mass.: Harvard University Press.

Ekman, P., et al. 2003. *Emotions Inside Out.* New York: New York Academy of Sciences.

Fausto-Sterling, A. 1985. *Myths of Gender: Biological Theories about Women and Men.* New York: Basic Books.

Fitch, W. T. 2006. "The biology and evolution of music: A comparative perspective." *Cognition* 100 (1): 173–215.

Gärdenfors, P. 2004. "Cooperation and the evolution of symbolic communication." In D. K. Oller et al. *Evolution of Communication Systems.* Cambridge, Mass.: MIT Press.

Gentner, T. Q., et al. 2006. "Recursive syntactic pattern learning by songbirds." *Nature* 440: 1204–07.

Haig, D. 1996. Personal communication with author and "Altercation of generations: Genetic conflicts of pregnancy." *Am J Reprod Immunol* 35 (3): 226–32.

Hinde, R. A. 1974. *Biological Bases of Human Social Behavior.* New York: McGraw-Hill.

Huron, D. 2006. *Sweet Anticipation: Music and the Psychology of Expectation.* Cambridge, Mass.: MIT Press.

Hyde, J. S. 2005. "The gender similarities hypothesis." *American Psychol* 60 (6): 581–92.

Kluger, J., et al. 2006. "How to spot a liar." *Time*, August 20.

Koelsch, S., et al. 2005. "Towards a neural basis of music perception." *Trends Cog Sci* 9 (12): 578–84.

Kuhl, P. K. 2004. "Early language acquisition: Cracking the speech code." *Nat Rev/Neurosci* 5: 831–43.

Matsumoto-Oda, A., et al. 2005. " 'Intentional' control of sound production found in leaf-clipping display of Mahale chimpanzees." *J Ethol* 23: 109–12.

Mehl, M. R., et al. 2007. "Are women really more talkative than men?" *Science* 317: 82.

Mithen, S. 2006. *The Singing Neanderthals.* Cambridge, Mass.: Harvard University Press.

Owings, D. H., et al. 1998. *Animal Vocal Communication: A New Approach.*" Cambridge, UK: Cambridge University Press.

Pika, S., et al, 2005. "The gestural communication of apes." *Gesture* 5 (1/2): 41–56.

Proverbio, A. M., et al. 2006. "Gender differences in hemispheric asymmetry for face processing." *BMC Neurosci* 7: 44–54.

Searcy, W. A., et al. 2005. *The Evolution of Animal Communication.* Princeton, N.J.: Princeton University Press.

Segerstråle, U., et al. 1997. *Nonverbal Communication: Where Nature Meets Culture.* Mahwah, N.J.: Lawrence Erlbaum Associates.

Slobodchikoff, C. N., et al. 2006. "Acoustic structures in the alarm calls of Gunnison's prairie dogs." *J Acoust Soc Am* 119 (5): 3153–60.

Thiessen, E. D. 2005. "Infant-directed speech facilitates word segmentation." *Infancy* 7 (1): 53–71.

10. TOUGH AS A BOILED OWL: PREDATORS

Bäckhed, F., et al. 2005. "Host-bacterial mutualism in the human intestine." *Science* 307: 1915–20.

Berger, L. R., 2006. "Brief communication: Predatory bird damage to the Taung type-skull of *Australopithecus africanus* Dart 1925." *Am J Phys Anthropol* 131: 166–68.

Chippaux, J. P. 1998. "Snake-bites: Appraisal of the global situation." *Bull World Health Organ* 76 (5): 515–24.

Crabb, P., et al. 2006. "Tool use increases survival of animal attacks: evidence for technological selection." Presented at Human Behavior and Evolution Society, Philadelphia, Penn.

Engel, C. 2002. *Wild Health: How Animals Keep Themselves Well and What We Can Learn From Them.* New York: Houghton Mifflin.

Floyd, T. 1999. "Bear-inflicted human injury and fatality." *Wilderness and Environ Med* 10: 75–87.

Hart, D., et al. 2005. *Man the Hunted.* New York: Westview Press.

Karanth, K. U., et al. 2005. "Distribution and dynamics of tiger and prey populations in Maharashtra, India," http://www.savethetigerfund.org.

McGraw, W. S., et al. 2006. "Primate remains from African crowned eagle (*Stephanoaeus coronatus*) nests in Ivory Coast's Tai Forest: Implications for primate predation and early hominid taphonomy in South Africa." *Am J Phys Anthropol* 131: 151–65.

McMichael, A. J., et al. 2005. "Global climate change." In *Comparative Quantification of Health Risks: Global and Regional Burden of Diseases Attributable to Selected Major Risks*. Geneva: World Health Organization.

Nicholas, C. G., et al. 2003. "The burden of chronic disease." *Science* 302: 1921–22.

Packer, C., et al. 2005. "Lion attacks on humans in Tanzania." *Nature* 436: 927–28.

Sharma, S. K., et al. 2004. "Impact of snake bites and determinants of fatal outcomes in southeastern Nepal." *Am J Trop Med Hyg* 71 (2): 234–38.

Stinson, S., et al. 2000. *Human Biology*. Wilmington, Del.: Wiley-Liss.

Thorpe, S. K. S., et al. 2007. "Origin of human bipedalism as an adaptation for locomotion on flexible branches." *Science* 316: 1328–31.

Zimmer, C. 2000. *Parasite Rex*. New York: Free Press.

11. A BULL IN A CHINA SHOP: ECOSYSTEM IMPACTS

Allen, M. S., et al. 2001. "Pacific 'Babes' ": Issues in the origins and dispersal of Pacific pigs and the potential of mitochondrial DNA analysis." *Intl J Osteoarchaeol* 11: 4–13.

Best Foot Forward, 2002. "City limits: A resource flow and ecological footprint analysis of Greater London," http://www.citylimitslondon.com.

Blackburn, T. M., et al. 2004. "Avian extinction and mammalian introductions on Oceanic islands." *Science* 305: 1955–58.

Chew, S. C. 2001. *World Ecological Degradation: Accumulation, Urbanization, and Deforestation, 3000 B.C.–A.D. 2000*. Walnut Creek, Calif.: AltaMira Press.

Erickson, C. L. 2000. "The Lake Titicaca basin: A precolombian built landscape." In *Imperfect Balance*, ed. D. Lentz. New York: Columbia University Press.

———. 2003. "Agricultural landscapes as world heritage: Raised field agriculture in Bolivia and Peru." In *Managing Change*, ed. J. M. Teutonico, et al. Los Angeles: Getty Conservation Institute.

Frederickson, M. E., et al. 2005. "Devil's gardens bedeviled by ants." *Nature* 437: 495–6.

Girardet, H. 1996. "Giant Footprints," http://www.unep.org.

Heckenberger, M. J., et al. 2003. "Amazonia 1492: Pristine forest or cultural parkland?" *Science* 301: 1710–14

Kirch, P. V. 1996. "Late Holocene human-induced modifications to a central Polynesian island ecosystem." *Proc Natl Acad Sci U S A* 93: 5296–300.

Larson, G., et al. 2005. "Worldwide phylogeography of wild boar reveals multiple centers of pig domestication." *Science* 307: 1618–21.

Lin, C. H. 2005. "Seismicity increase after the construction of the world's tallest building: An active blind fault beneath the Taipei 1–1." *Geophys Res Lett* 32 (22): L2231.

Meharg, A. A., et al. 2003. "A pre-industrial source of dioxins and furans." *Nature* 421: 909–10.

Nelson, D. M., et al, 2006. "The influence of aridity and fire on Holocene prairie communities in the eastern prairie peninsula." *Ecology* 87 (10): 2523–36.

Ponting, C. 1992. *A Green History of the World: The Environment and the Collapse of Great Civilizations.* New York: St. Martin's Press.

Pyatt, F. B., et al. 2000. "An imperial legacy? An exploration of the environmental impact of ancient metal mining and smelting in southern Jordan." *J Archaeol Sci* 27: 771–78.

Pyne, S. J. 1997. *World Fire.* New York: Henry Holt.

Rich, C., et al. 2005. *Ecological Consequences of Artificial Night Lighting.* Washington, D.C.: Island Press.

Russell-Smith, J., et al. 1997. "Aboriginal resource utilization and fire management practice in Western Arnhem Land, monsoonal Northern Australia: Notes for prehistory, lessons for the future." *Hum Ecol* 25 (2): 159–94.

Savelle, J. M., et al. 2002. "Variability in Palaeoeskimo occupation of south-western Victoria Island, Arctic Canada: Causes and consequences. *World Archaeol* 33 (3): 508–22.

Shotyk, W., et al. 1998. "History of atmospheric lead deposition since 12,370 14C yr BP from a peat bog, Jura Mountains, Switzerland." *Science* 281: 1635–40.

Steadman, D. W., et al. 1990. "Prehistoric extinction of birds on Mangaia, Cook Islands, Polynesia." *Proc Natl Acad Sci U S A* 87: 9605–09.

Tanno, K., et al. 2006. "How fast was wild wheat domesticated?" *Science* 311 (5769): 1886.

Thery, I., et al. 1995. "First use of coal." *Nature* 373: 480–81.

Wilkinson, B. H. 2005. "Humans as geologic agents: A deep-time perspective." *Geology* 33 (3): 161–64.

Williams, T. M., et al. 2004. "Killer appetites: assessing the role of predators in ecological communities." *Ecology* 85 (12): 3373–84.

World Wildlife Fund. 2004. "Living planet report 2004," http://assets.panda.org/downloads/lpr2004.pdf.

Vos, D. J., et al. 2006. "Documentation of sea otters and birds as prey for killer whales." *Mar Mamm Sci* 22 (1): 201–05.

INDEX